PLASTIC

PLASTIC

AN AUTOBIOGRAPHY

ALLISON COBB

Nightboat Books
New York

Printed in the United States

ISBN: 978-1-64362-038-1

Cover design by Mary Austin Speaker
Cover photograph courtesy of Jerry Takigawa from the "False Food" series (2010). The plastic in the image was collected by scientists from the remains of albatross on Midway Island and loaned by the Monterey Bay Aquarium.
Design and typesetting by HR Hegnauer
Text set in Athelas and Requiem

Cataloging-in-publication data is available from the Library of Congress

Nightboat Books
New York
www.nightboat.org

CONTENTS

GIFT: THE THING

I. THE THING

II. REFUSE

III. THE LIVES

Knowing does not come from standing at a distance and representing but rather from a direct material engagement with the world.

—Karen Barad

GIFT:
THE THING

I wanted to trace the lost patterns that came before the world was broken and find the new ones we could make out of the shards.

—Rebecca Solnit

The thing turned up in a corner of the yard, just outside the fence. I found it when I went out to take Quincy for a walk. Curved and black, plastic. Four feet long, a foot at its widest. I thought at first it was a car bumper. I put it in the grass by the porch. The next morning it was still there. I sat next to it in the sun and looked closely.

It was not the first piece of plastic junk I had sat staring at. For nearly a year, I'd been picking up all the plastic I found on my daily dog walk. I'd been arranging it into patterns, taking photographs. I'd been storing it all up in plastic garbage bags on the back porch.

I didn't know exactly why I was doing this. I wanted to understand something. Plastic on the dog walk. Plastic on visits to the beach. Plastic studding the ground, everywhere I looked. I gathered it all up.

I am the *no* and the *yes*—

A line from the poet 'Annah Sobelman's first book. It has lived with me for years, sometimes whispering through my mind in its old remembered rhythm. In the poem, Sobelman follows the line with a qualifying phrase. She narrows it, makes it domestic, but I want the raw declaration, hanging there on the turn of itself:

I am the *no*

and the *yes*

For nearly half my life, I have worked for an environmental group. I spend most of my days in front of a computer screen, taking in a deluge of information about planetary trauma and emergency.

Most of it floods through me—too vast to grasp. But plastic was a shard that stuck. Plastic I could touch, and it could touch me back.

On this day, as I sat beside the car part, I was thinking about the German philosopher Martin Heidegger, his essay called "The Thing." His book *Poetry, Language, Thought* had sat on my shelf, unopened, for more than a decade since graduate school. A few days earlier, on a whim, I picked it up.

Heidegger writes that distance disappears and all things come equally close because of technology. In 1949, when he wrote the essay, he meant airplanes, the radio, TV. These inventions bring everything before us in image and sound: ancient Egyptian pyramids, a cat in Japan, glaciers shearing off into Arctic waters. Everything flattens out—a "uniform distanceless," as he calls it.

But this does not make anything present. The only way to approach a thing, to bring it near, is by sidling up to it, by thinking around, or through, what appears obvious. He performs this kind of meditation on the "thingness" of a clay jug.

I thought of meditating on this cart part. It had a smooth surface, shiny. One side was flecked with light splash marks from mud or paint. It formed a complicated shape—wide at one end and tapered at the other, with holes and slats and ridges all along its length. The widest end contained deep score marks, some scratched all the way through the plastic. It was stiff but still pliant. Without the car body to hold its curved shape, it folded in half like a wing at its narrowest point.

The image that came to me was an albatross carcass bursting with plastic. This was the first shard that stuck, a snippet in a news story about a piece of plastic from World War II, found inside a dead albatross chick sixty years later. It stayed in my mind for years, like the *no* and the *yes*.

I dragged the car part inside the house. Nearly ten years later it sits beside me, near my desk. I learned this: that the world is not broken. Or that it has always been shards, kaleidoscopically interwoven, not one world, many, threaded through one another, like fungus hyphae through soil.

Worlds end. As Kathryn Yusoff points out in *A Billion Black Anthropocenes or None*, some worlds have ended over and over, lives consumed and discarded by individuals woven into systems that give them life-and-death power. Like settler colonialism, like capitalism, these are systems built by humans, but they exceed individuals. They extend across generations and geographies—planet-scale forces of destruction.

Plastic waste stems from this consume-and-dispose violence. I learned that waste is not an unintended consequence of a miracle new technology. Waste is inherent in plastic production as it accelerated after World War II.

In 1945, days before the U.S. military incinerated two cities with atomic bombs, a DuPont executive looked forward to the end of the war and the surge of buying that would follow as soldiers returned home and bought houses and cars, washing machines and refrigerators. The job ahead, he told a group of marketing experts: "See to it that Americans are never satisfied."

Plastic embodies this infinite desire. Conjured out of gas and oil—the seemingly endless reservoirs of dead plants and animals underlying Earth—plastic transmutes death into eternal life. The word plastic refers not to any specific object, just the quality of a material—that it is capable of taking shape, an endless stream of shapes.

Objects formed from plastic ease suffering and save lives—artificial hearts, IV bags, the tubes snaking out of a respirator. Plastic makes cars safer, airplanes lighter, and delivers drinking water. The single largest use of plastic, though, is for containing other objects. Forty percent of plastic goes into packaging, to be used once and then discarded, driving endless demand for more.

Companies work to keep these facts hidden. When the evidence becomes too overwhelming—plastic clogging roadsides, oceans, living bodies—companies shift responsibility onto individuals through things like anti-littering campaigns, and ensure that taxpayers and municipalities pay the tab for managing the waste. The lives harmed at every step—human and nonhuman—drop out of the equation.

The same consume-and-dispose violence threads through me also. It has benefited me my whole life. I grew up the daughter of a nuclear physicist in Los Alamos, the town that built the atomic bombs, which ended some lives in order to save others perceived to have more value. We are woven into the same net—me, and bombs, and this car part.

For a decade, I followed threads that tie us together, through airplanes and sailors, the hydrogen bomb, Pacific islands, the Nazis, and Heidegger. I followed threads through silence, loss, and grief, through the birth of chemistry and the invention of radar, through patriarchy, empire, and chattel slavery. I followed threads through apologies and their failure, through a pandemic, and an uprising, and living lungs struggling to breathe, through old wounds and new ones, hurt reverberating, aching to be re-membered.

This object, a book and its journeys, this broken-down car part, its life. This is my *no.* I have wanted this car part and its entanglements—often ugly ones, and painful—to leave me. I have wanted not to have to face, in my privilege, the terms of its existence. I learned this: There is nowhere to go. The same terms that made this piece of plastic made me, my own body, and each of my breaths. This is also—it must be—my *yes*.

I.

THE THING

WORK

I found a way to make it work.

That's what Stan Ulam told his wife, Françoise, one bright winter noon in 1951, when she walked into the living room of their house in Los Alamos and found him staring out the window.

"What work?"

The mathematician turned his eyes to her.

His eyes.

I have tried to figure out what they were like. I have studied the photos, most black-and-white. So flat. They just signal dark and light.

His friend, the mathematician Gian-Carlo Rota, said what they were like: "He looked at me, his intense blue-green eyes popping and slightly twitching (they were the eyes of a prophet, like Madame Blavatsky's)."

The word prophet is ancient Greek: one who delivers messages from gods. Medieval Christian writers suppressed the Latin word for prophet because it breathed a pagan breath. They inscribed instead the safely dead Greek.

The Latin they reserved for poets. *Vates.* It comes from an ancient root that means "to blow into, inspire." As in, entered by a spirit. The name for the Norse god Odin sprung from this root, as did the Old English word "wood," which means insane.

What has died—a language, a person—can be shaped into anything. A mask to fit the purpose. Is that accurate?

I meant to write something brisk, concise. But this has taken me far from Stan Ulam's eyes.

I could write to Claire, Ulam's only child. She would be about my parents' age. Born of the war. She and I share a birthplace—Los Alamos. We grew up in different generations, but both on the edges of "the Lab," daughters of nuclear fathers.

I could start with that. But what would I ask? A single question, concerning his eyes?

What can be said: He turned his eyes to her, Françoise, his wife.

This is her story. She typed it into the back of his autobiography as a postscript to his *Adventures*. She called it "engraved on my memory."

No. It can't even be said he looked at her. She didn't say so. I imagined it. Maybe he continued to stare, as she wrote, "unseeing," out the window.

"I found a way to make it work."

"What work?"

"The Super," he replied, by which he meant the new bomb, thermonuclear. "It is a totally different scheme, and it will change the course of history."

STORY

Ensign Elwyn Christman piloted his airplane through a moonless night over the ocean, the cockpit illuminated by the phosphorescent hands of the instruments and the flickering light of the exhaust stacks. He hunched inside his wool flight suit against the cold at twelve thousand feet.

The date: December 27, 1941. Christman. The name so filled with portent. Then, Elwyn. It is Gaelic, or Welsh. It means white and fair, holy. He went by Chris to his friends. To his mother he remained Elwyn.

On this night, Christman's commanders in the VP-101 Navy squadron ordered him and his crew to fly eight hundred miles and bomb Japanese warships at Jolo Island in the Philippines. They flew in formation with two other planes. Another group of three flew nearby, all without navigation lights, invisible in the dark.

I know what it was like inside the cockpit because of another man on that mission: Joseph F. Long. His nephew saw my query on a veterans' listserv and emailed his uncle's journal entries.

Long's handwriting is looping, elegant. He wrote: "The cockpit is illuminated by the phosphorescent hands of the instruments and the flickering light of the exhaust stacks." He wrote: "No moon," and "The big dipper is upside down off our starboard bow."

Long disappeared on a mission two months later.

War leaves mostly gaps. But that sounds too clean, doesn't it? Raw, jagged, taking blood from each birth that has to pass.

How to cross to Christman, a once-living human?

More than others, he left a trace. Because he was a leader, a Lieutenant Commander in the Navy in charge of his own airplane, and later, an entire squadron. And because of this mission, which they told, and others retold, so it became a story.

The PBY airplanes that they flew they called "lumbering," slow and heavy. Meant for patrol, not combat. They were supposed to be escorted by fighters. But here is history: On December 7, 1941, the Japanese military destroyed nearly the entire U.S. fleet at Pearl Harbor then moved on, wiping out bases across the Pacific, forcing the survivors back fifteen hundred miles across the ocean. There were no fighters left. Of the six planes sent on this mission, two would come back.

I have Christman's own words about what happened next, typewritten as an official report to the Commander-in-Chief, U.S. ASIATIC FLEET. I received it in the mail from the federal government, for free, as the letter told me, "due to the small number of pages."

I have no way to know whether Christman typed the report himself, but I can picture him in rumpled khakis in a sweaty, tropical barracks, pounding on an old typewriter. The image comes in part from black-and-white World War II movies, in part from photos sent to me by his son. Christman: dark hair slicked back, dark eyes, deep cleft in chin, dark mustache, age twenty-six.

Whoever typed the report made several errors and didn't bother to fix them. The idiosyncrasies of the machine left their mark also—the B has a hitch in its curve, which makes the word "bomb" waver on the page. Christman kept emotions out of his account; he reported just the facts.

He brought his plane over Jolo just at dawn. He didn't realize they were under attack until he heard his machine gunner firing. Japanese fighters in Zero airplanes. Christman zigzagged, trying to avoid them. He watched bullets pour into the cockpit, smashing the instrument panel. I imagine this detail in slow motion, a movie, every sense narrowed to that.

The Zero pilot came close and fired his 20-millimeter cannon. The shell entered the cockpit just above Christman's head. He ducked, he said, by instinct. The cannon punctured the fuel tanks in the wings, and gasoline poured into the plane. A second cannon shell exploded, igniting the fuel. Christman's copilot turned to look back into the plane. All he could see were flames.

ZERO

The ancient Greeks rejected the concept of zero. When they looked at the universe, they saw numbers. The big secret that Pythagoras hoarded, the key to it all, resided in numbers' relation to one another. Ratios. Greeks found the same ratios repeated everywhere: in music, in planetary motions, in flowers. They considered this beauty (all things poised in a balance) and worshiped it. But ratios cannot be made with zero. Zero consumes all relations

and spits out

what

Chaos

which for the Greeks signified a kind

of absence

without form

or limit—

A weak word,

Chaos, it comes from gape,

just the opening of the human mouth. Zero,

weak also, just

empty place, then

wind, nothing.

NOTHING

What can follow?

It silenced me for some time.

I ended with nothing, then found I could not put a mark in the white space that followed.

Try this: dwell a bit in that gap. *I rang the doorbell of nothing there* —Alice Notley.

No—

The tiny word intertwines two roots. First the "n" sound *ne, na.* It seemed always to signify absence, a lack. Perhaps that's the tongue pushing out from the mouth. Away, not here. Then the round-mouthed *aiw*: vital force, long life. It echoes also through *aye* and *yea*.

I am the *no*

and the *yes*. The no

-thing desire

toward being. So

this: the dead albatross chick

on its back in the grass. A knife slides

inside its white feathers. *Mercy floated towards me, towards*

the door not there.

The cut reveals a black cylinder of plastic, a bottle cap, a toy top.

Plastic erupts from inside the chick.

THE IMPURITY

1856. An eighteen-year-old stood before a table crowded with beakers. He had thick black hair that wanted to spring into curls, but he carefully slicked it down against his skull. He had set up a makeshift lab on the top floor of his family's small house in London's East End. He was spending his Easter holiday trying to do something miraculous with chemistry, to distill from black coal tar a white, crystalline powder: quinine, the only known cure for malaria.

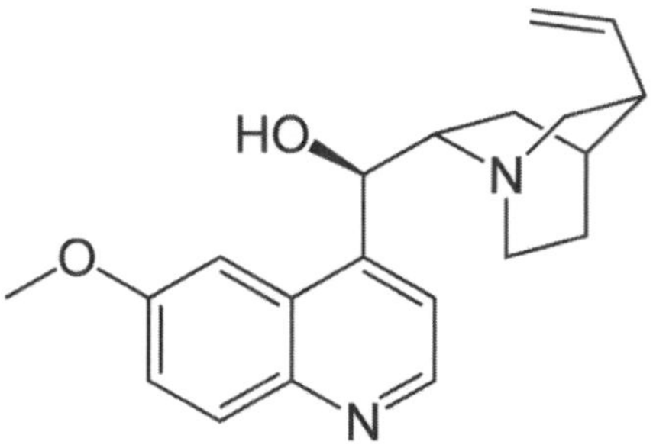

William Perkin heated and condensed, mixed, and extracted. He ended up with the opposite of white powder, a substance sticky and black. He rinsed the gum from his flask with alcohol and noticed something else, a color. Luminous purple. He tried dipping a square of silk in his solution and saw that it was beautiful.

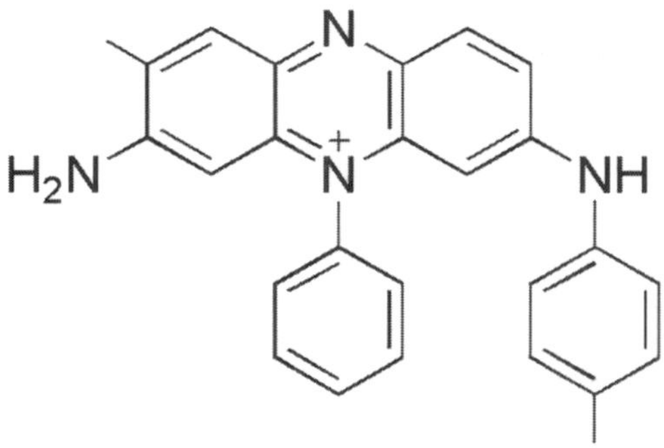

The richness of the color came from what is called an impurity. That means Perkin didn't know it was there. At the time, chemists lived in this state of not knowing. They couldn't see the molecules they tried to manipulate, and they didn't understand their structure or the forces that held them together. The word molecule reflects this lack. It just means something vague, a tiny mass.

Perkin got his chemicals from the oily black tar left by heating coal to make the gas that burned in lamps across London. Chemists all over Europe conducted experiments to see what they could make from this stinking waste. They knew it contained a complicated mix of chemicals they might be able to turn into something useful. Plus, it was free for the taking. Most people considered chemistry an esoteric pursuit. Industry consisted of steam-powered metal, not strange concoctions from a laboratory. Perkin's father, a carpenter, wanted him to study something more practical.

Reactions with components of coal tar often bled colors—crimson, blue, pale yellow—just shifts in the wavelengths of light. Chemists ignored them. They suggested no practical purpose. But Perkin was a painter. He had an eye for color. And he was a beginner, still practically a child, naive enough to pursue something with no utility.

Perkin's discovery would launch a race for molecular plunder and spark industrial chemistry. He had no idea of that future as he stood in his parents' house, holding his scrap of purple silk. He decided to name the color Tyrian purple after the ancient Phoenician city. That city, Tyre, produced a purple dye prized by the elite of different cultures over a thousand years, from Egyptian pharaohs to Roman emperors.

The original Tyrian purple came from the mucus of sea snails. Each animal produced a few grey-yellow drops that turned purple on exposure to the sun. It took twelve thousand snails to make 1.5 grams of dye, enough to color a single robe. The ancients rated purple with gold and silver—only the wealthy, royalty, and priests could wear it. *The holy purple mollusk must be worshiped*, wrote Theodosius.

The Phoenicians wiped out two species of purple mollusk; two others still exist. They have delicate spiral shells, the kind beach collectors like. Evidence still exists of the ancient frenzy for purple: smashed shells several feet thick cover the beaches around modern-day Tyre.

WORK

In my work for an environmental organization, I am paid to create certain feelings in other people. I use words. With words I convey, which means "to carry along." I act as ghost of the present, the past, and the yet to come. I conjure for others the harm and suffering that humans cause to the world. The people who can't breathe because of pollution. The babies born with hundreds of industrial chemicals circulating through their bloodstreams. The beings that have ceased to exist, wiped out by human actions. And the worse that is still in store.

This is the job I do: to make others feel sorrow and concern, but not despair. To give them hope, which means "to desire and expect," to look forward with surety that the thing they want will come to pass.

The hope I give is exact: That privileged people can stem the tide of damage, waste, and plunder embedded in our global economy without changing the terms of the system that benefits them, and me also.

This is mercy, which means forgiveness for offenses, but comes from the Latin word for "fee, or bribe, wages."

I earn my wage performing hope for others. This requires me not to feel myself. Not to lie on the floor howling, crying out, weeping.

This suits me.

I learned very young how to survive in the world, what is required of me. I learned this by watching: that men act. Men have power. They don't feel, except for anger, which engenders fear and increases their power.

Women feel. This work puts them at risk. It makes them unseemly. Weak. Without intellect. I watched, and over time, I grew a sense of having a wide, lidless eyeball inside me. I became only mind and eyeball. I watched, and considered how to exist.

I remember being told I had my father's eyes. I wept because I imagined him away at work, unable to see. Or maybe I wept because the eye is what I had left, and it wasn't even really mine after all. I remember riding in the backseat, staying silent for so long that my dry lips sealed together. I remember not wanting to split them to speak.

Now I speak constantly. For hours a day, words flow from me. Some of the words bring me money. The writing I do for no wage helps me to think, to figure out what I know, and how to continue. Those words become books. But behind all of these words is a silence. A gap. The word comes from Old Norse: it means "chasm, empty space."

In this formlessness resides someone else. A being who might be intact. Who might have the skill to feel as well as think and act. Mostly this one stays trapped, walled off. Survival requires this. The watcher believes it. Dear eyeball, keeping safe.

I am the *no.*

I prefer numbness, the uniform distanceless. But *no* is a precipice, hanging on the turn toward *yes.*

The hidden one defies her trap. She has beast in her, and fire, and light. She wants the outside to touch her. She wants to touch back. Let the shards of worlds catch.

THE MARINER

Heidegger's essay "The Thing" startles me. He refers to the atomic bomb, and asks a strange question: "Is not this merging of everything into the distanceless more unearthly than everything bursting apart?

"Man stares at what the explosion of the atom bomb could bring with it. He does not see that the atom bomb and its explosion are the mere final emission of what has long since taken place, has already happened. Not to mention the single hydrogen bomb, whose triggering...might be enough to snuff out all life on Earth. What is this helpless anxiety still waiting for, if the terrible has already happened?"

: :

Once I heard my mother's father tell my father he felt grateful to all the people working at Los Alamos. He said he was glad for the atomic bomb because it meant he didn't have to invade Japan. He stood in his sailor blues and watched from the deck of a ship in the Bay of Tokyo the signing of the Japanese Instrument of Surrender.

My grandfather enlisted in the Navy as an Apprentice Seaman on May 12, 1944, at the Empire Building in Rockford, Illinois. He was twenty-four. He had married my grandmother—whom he had known since grade school—three years before. They had one child, my mother, who was a little over one year old at the time. He joined the Navy before he could be drafted into the Army because he thought he had a better chance of surviving at sea. That's the family story.

In the "Excerpts from Physical Exam recorded on the day of enlistment," he answered "No" to the following questions:

Spells of unconsciousness?

Nervous trouble?

Convulsions?

Asthma?

Hay fever?

Enuresis?

He had never been treated in a hospital, asylum, or sanitarium. His complexion was ruddy, hair blond, eyes blue. He was six feet tall and one hundred ninety pounds. His appearance was "GOOD" and everything else about him "NORMAL." His blood Kahn test (for syphilis) was negative. His chest x-ray was "essentially negative."

I paused at this, and I wonder if he did also. What did "essentially" mean? What little whisper or glitch of no consequence existed inside his chest?

I received my grandfather's war records in the mail from the National Archives. It cost seventy dollars and took several months. The brief letter accompanying the packet arrived with a coffee stain in the upper left corner. The letter states: "We regret if the photocopy is of poor quality; however, it is the best we can obtain."

The official record of my grandfather's war service has nothing to distinguish it. He received no medals. For nine months, he crisscrossed the U.S. from Illinois to Virginia to San Diego. He received recruit training, a basic course in engineering, and qualified as a motor machinist. He passed landing craft school, which equipped him to pilot small boats dropping off troops and authorized him to wear "amphibious force insignia."

At each step, he received a rating card. His mark for "Conduct" was always 4.0. For his final ranking, the Executive Officer gave him a 3.7 for "Proficiency," a 3.9 for "Mechanical Ability," and a 3.5 in "Ability as Leader of Men."

On January 25, 1945, he embarked for the Pacific aboard the USS Cullman. He spent the rest of the war on the water.

: :

The poet Samuel Taylor Coleridge was not a sailor, and he never saw an albatross. William Wordsworth claimed he gave his friend the idea for *The Rime of the Ancient Mariner* after reading a journal of sailing around the world by Captain George Shelvocke. The captain recorded that one of his sailors shot a black albatross that for days had followed the ship.

When the sailor kills the bird in the poem, every vital force seems to fade. The winds stop, the ship stands still, the world falls silent. Only a bloody sun burns. The others hang the albatross around the sailor's neck, but the real punishment is this: The mariner lives while everyone around him dies.

> Each turned his face with a ghastly pang,
>
> And cursed me with his eye.

The ship becomes a world of corpses. Even the sea rots. All that breathes is monstrous.

> And a thousand thousand slimy things
>
> Lived on; and so did I.

Coleridge published his poem about a hundred years before Heidegger was born. When Heidegger looked around at the mid-twentieth century, he described something like the mariner's nightmare: corpses in a world of corpses. He believed that technology had distorted humanity's relationship to all that exists, and reduced everything (objects, animals, people, even oneself) to raw material, a stockpile of useful traits: labor power, money value, explosive force. As a result, Heidegger thought, everything dies. That's why he saw the hydrogen bomb as "mere," only the logical extension of this state.

: :

In the car on the way home from my grandfather's funeral, my aunt revealed the following: When she was a teenager, a man came to visit who had served in the Navy with my grandfather. The man said that my grandfather saved his life. They somehow got separated from their ship and marooned on an island. Two Japanese soldiers attacked. Armed with nothing but a knife, my grandfather killed them both.

Norm. My grandfather: Norman Isadore Olson. Son of Swedish immigrants. He worked his entire life after the war at Woodward Governor, a factory that made airplane parts. The governor regulates the speed and power of a motor. My grandmother once said that when he got back from the war he had nightmares, screams yanking them both from sinking sleep. She said this as if it were something dark, a secret. He told her only that it was hard to spend all day inside the factory after so long on the open water.

I asked him once to tell me about the war. He told me this: His bunkmate on the ship always ate his ice cream first because, as he would say, life is short.

What I remember: He was enormous, tall and broad, with a high forehead and a combed back sweep of grey hair. Blue eyes that sparkled. My grandparents lived far away in Rockford, Illinois, where my parents grew up, which my mom still called "home." We saw them once a year. He may have represented for me a different possibility for being male, but I don't think I understood him like that. He was just a grandfather. Sweet and gentle. I played with his hair and he chuckled and grinned at me. His breath smelled like coffee. Once, he snuck my sister and me out for an early morning boat ride and pancakes at the diner.

He liked most of all to sing: popular songs from his day and old spirituals like "Swing Low, Sweet Chariot," and "Were You There (When They Crucified My Lord)."

Sometimes it causes me to tremble, tremble, tremble

He cut a few records with some buddies back in the '40s. My uncle found them in the trunk of my grandfather's old car. No one wanted them but me. They were scratched and brittle, pre-vinyl, the black shellac in places peeling off. I gave them to my sister, who is a musician, hoping she might find some way to translate them into a medium that would let us listen. At some point, she or my mother threw them out.

It has been nearly two decades since my aunt told her story, and it all happened in the haze of post-funeral. My mother doesn't remember it, which makes me wonder if my aunt really did tell us. I'm afraid to ask her now, just like they were afraid to ask him then.

We won't get them back. They are the silence we cannot break. My grandfather, gone where the dead go. Into ground, certainly. The records. I can picture them, grooved surfaces melting off beneath a mountain of junk at the Los Alamos County Dump.

THE CURTAIN

September 1, 1939. Stanislaw Ulam lay sweating in a Columbus Circle hotel room on a humid New York City night. He couldn't sleep. Around one or two in the morning, the phone rang. It was his friend Witold Hurewicz, a fellow Polish mathematician. Ulam remembered the quality of Hurewicz's voice, somber and throaty. "Warsaw has been bombed," he reported. "The war has begun." Ulam decided not to wake his sixteen-year-old brother, Adam, sleeping in the bed beside him.

They had not seen war coming. It was a "practice of magical thinking," Adam would later write, that convinced most ordinary Poles and their leaders that Hitler would not dare attack Poland even as he swept through Austria and Czechoslovakia.

The Ulam brothers came from a prosperous family of lawyers and bankers in Lwów, ancient meeting point of East and West, crowded with stone buildings and cafes filled with people at all hours. Stan's parents encouraged him to get a practical education, so he studied engineering at the Lwów Polytechnic School, but soon the mysteries of math drew him in. He established an early reputation as a brilliant mathematician. He spent his time outside of school in coffee shops with professors and fellow students, working through the latest problems in the field. He claimed that one such session lasted seventeen hours.

Stan earned his doctorate in 1933. Few positions for mathematicians existed in Polish universities, and, as a Jew, Stan knew he had no future there anyway. Through friends, he secured appointments at Princeton and then Harvard, but these were only temporary. Stan formed a particle in a flood of European intellectuals seeking refuge at U.S. universities. The American elite did not always welcome them, and in the midst of the Depression had few openings to offer. So Stan taught at various U.S. institutions throughout the 1930s, and went home to Lwów every summer.

In 1938, the Ulams' mother died. No one records much about her, except that she had a long illness requiring many surgeries. The family crisis unfolded inside one of national proportions. Workers ravaged by the Depression rioted in the streets; the weakened regime courted the Fascists and Nazis to shore up its power; waves of anti-Semitic attacks swept the cities. Adam compared the change in atmosphere to being suddenly transported to a high altitude—one had to gasp for air.

They decided Adam, who was thirteen years younger, should go with Stan to study in America. Brown University accepted him for the fall term, 1939. The brothers spent the summer as they always did, lounging in cafes with friends. They took a trip to their uncle's estate in the mountains. They planned to leave September 3.

Only their father seemed to sense what was coming. He pleaded with them to go earlier. But even he could not envision the scope of the coming catastrophe. He made financial arrangements for Adam for one year, expecting them both to return the next summer.

In mid-August, Stan, Adam, their father, and their uncle took the six-hour train ride from Lwów across Poland to the Baltic port of Gdynia. Stan and Adam boarded the MS *Batory* bound for the U.S. It would be the last ship to leave port.

A few days later, a German battleship entered the same harbor with two hundred twenty-five stormtroopers on board. September 1: German forces occupied the Polish seacoast and swept south through the country.

Date, from the Latin for "given," now takes

its deeper

dark. Framed

in flame—

a scar, each

flesh event

Stan and Adam never saw their family again. For the most part, they never learned what happened—a few agonizing clues over the years: their sister and her baby probably shot by Nazis; their father last known sheltering a refugee boy about Adam's age in Lwów, burning his law books to keep warm.

"The society in which [Ulam] grew up and was educated was almost completely obliterated during World War II," wrote the Polish American mathematician Jan Mycielski.

Of that middle-of-the-night phone call, Stan later recorded: "It was as if a curtain had fallen on my past life, cutting it off from my future. There has been a different color and meaning to everything ever since."

CAR PART

Heidegger would not consider the car part, an object of technology, worthy of meditation. I decided to try it anyway. From studying its shape and looking at my own car, I determined it is a liner for the inside of the fender. The word fender is short for "defender," and it first referred to a piece of woven rope that kept the hull of a ship from smashing against the wharf. People added fenders to early cars because fast-turning wheels tossed up dirt and stones hazardous to other cars and people.

The fender liner acts like a shield. It keeps dirt and water out of the engine and deadens the impact of stones striking the body. It is curved like that, shaped like a shield, for protecting. It has an aesthetic function also. Matte black, it gives the base of the car a finished look, maybe not something you even notice as your eye runs over it, just a smooth surface, something pleasing, complete.

It fulfills, in a small and specific way, two human needs, this piece of plastic: for protection, and for beauty. Not the beauty that Pythagoras admired, arising from a harmony of parts. This is beauty as a mask for something ugly. It covers the car's raw metal underbelly, which betrays its brute machine birth and pierces the illusion of speed and ease the shining surfaces impart. Plastic, the perfect cover: indeterminate, no whiff of industrial blood on it, featureless, flawless, eternal. It draws into itself each object, each

instant formed to

sate such

ancient

lust: to guard,

and soothe,

protect. Or,

more basic, just

that longed-for

state:

a peace

at rest, accordant.

This requires

defense against

some

Other.

Right?

FRAGMENT

Few tourists visit Kamilo, which means "swirling currents" in Hawaiian. The only way to get to this beach is to drive five miles on a fading dirt track over lava, rock, and sand. We asked Noni Sanford to guide us. She is a local artist I read about in a book on ocean trash.

Jen and I traveled to Hawai'i in the spring of 2011 as a gift from my family. We expended about 2,500 pounds of planet-warming gases to get there. We met my parents, my sister, her husband, and their children on the Big Island. Jen and I—in our years together, we had not found a word to describe our relationship to outsiders that didn't feel foreign, jarring. We needed no word for one another, in the intimacy of our lives together. To represent ourselves to others, I suppose we used "partners." Not "wife," ugh—besides, marriage for us then was not legal. Partner was probably the word I used when I spoke with Noni Sanford on the phone. I told her Jen and I would meet her and her husband, Ron, beside the road at Wai'ōhinu Park, near the island's southernmost point. Noni gave precise orders. Rent a four-wheel-drive, two-door Jeep, and do NOT tell the rental company where you plan to go.

We drove through early morning mists, ticking off mile markers on the map. At the park, Noni and Ron stood dwarfed by their Unimog, a bright orange truck with huge tires. I learned later that Mercedes first built this vehicle for driving around war-ruined Germany. It roughly stands for "universal motorized implement machine." This is how Noni and Ron get from their home on the smoking flanks of Kīlauea to the roadless beach at the tip of Hawai'i, youngest island in the chain.

We approached each other, four strangers. Noni had a grey braid that hung all the way to her waist, high cheekbones, and dark eyes outlined with black liner. Ron was tall and fair, with the sun- and wind-burned look of a person who spends a lot of time outdoors. Noni opened her arms to me. "I'm sick," I warned, leaning away. "I woke up with a sore throat." She rolled her eyes, pulled me in.

It took an hour to go the five miles, Noni at the wheel of our green Jeep, moving at a crawl behind Ron, tires grinding over boulders, across ditches and drop-offs. I sat crammed in the back with two scientists who were in Hawaiʻi for the Fifth International Marine Debris Conference, who had also contacted Noni to guide them to Kamilo. We grasped the seats, trying not to fall into each other's laps when the Jeep tilted sideways on a rock or tipped vertically into a pit. But we made it. Noni nicknamed the truck Greeny Guy. We all cheered for it.

This is what it looks like at Kamilo: black lava spills right to the seething water. Native naupaka shrubs and low heliotrope spread green across thin stretches of sandy soil. Mauna Loa, Earth's largest volcano, looms dark and purple. It erupted in 1868 and spewed the molten rock that streams across this beach.

We picked our way over broken flows. A misstep and jagged edges cut into skin. Ron clocked the wind with his handheld meter: forty miles an hour. We couldn't hear one another speak unless we stood quite close. The wind tore the sound from our mouths and flung it down the beach. It threw stinging sand and salt at our skin. Hats, even tied, wouldn't stay on. Sunglasses and camera lenses fogged with salt spray. The sun beat down.

Noni led us forward toward the water. My feet began to slip. Beneath us was no longer only the wicked-looking lava. The tides had tossed up a whole cosmos: tree trunks, coconut shells, coral, car tires, barrels, and plastic. Bright colored plastic of every shape and size shifted and slid beneath our feet.

In high places I get vertigo. Instinct pulls me to my hands and knees, as close to the ground as possible. A similar thing happened at Kamilo—wind, sun, rock, water. I squatted and looked straight down at the plastic ground. I dug into it with my fingers, burrowing toward—what—solid earth?

I didn't know what else to do. The watcher inside me took over, and I began to make a list of all the plastic I could identify:

Styrofoam, large chunks, like rocks strewn across the beach

plastic wedge heels for platform shoes, oddly common—I saw four of them

fishing floats, all shapes and sizes, with characters in Japanese and Korean

Febreze bottle: "Eliminates odor and freshens the air. Spring + renewal scent."

Epson printer cartridge, barely hanging together

motorcycle helmet, purple lining encrusted with barnacles, sitting next to a brown hard hat

so many plastic bottles they would be impossible to list: some looked new, like they just came from the store, most weathered, many with Japanese characters

bottle caps, so common they faded into backdrop

brown toy dog with blue collar, missing its tail

blue, black, and white striped basketball

motor oil bottle

cigarette pack

brand new tennis shoe

another motorcycle helmet

hiking boot

foam sheet with flip-flop shape cut out

white jug with Japanese characters, in English: "Antisepsis Remove Bacilli, The Blanch Water"

two pink combs adorned with flowers

black combs, several

three umbrella handles (Noni collects these)

perhaps a dozen disposable lighters

three golf balls

a blob that Noni identified as the remains of a superball

one jack

globs of a whitish waxy substance

partly incinerated twisted hunks

a chunk of tiny reflective yellow beads Ron said are melted and used for road markings

two round chewing gum containers with Japanese characters; in English one said "Xylicube Lime mint," the other "Fruits Gum"

plastic structure for the heel of a shoe, which formed an "A," so I kept it

brown plastic shell Ron identified as half an octopus trap

scrap of Astroturf

Noni had disappeared down the beach. Ron seemed worried. We pulled our eyes away from the junk—it was kaleidoscopic, mesmerizing—and gathered around him in a little windblown clot. He said he was going to take the truck and go look, but then she appeared, first her head coming over a rise, and then the rest of her, the bag she made out of an old t-shirt slung over her shoulder. She showed us her find: a pristine white plastic shard that read "GOD."

THE RING

1862. August Kekulé, a thirty-two-year-old professor at Ghent University in Belgium, was sitting in his study struggling to finish writing a textbook on chemistry. He dozed off in front of the fire. "The atoms fluttered before my eyes...everything in motion, twisting and turning like snakes. But look, what was that? One of the snakes had seized its own tail, and the figure whirled mockingly before my eyes. As if by a flash of lightning I awoke..."

Kekulé had envisioned a ring-shape for benzene. The clear, sweet-smelling liquid had puzzled chemists for thirty years since Michael Faraday first separated it from whale oil. Benzene remained stable, combining with few other substances. When it did react it behaved unlike anything else.

Kekulé's vision, one chemist wrote later, made sense of existing events and threw a flood of light into the future.

Kekulé did not publish his discovery for some years. In the meantime, he married Stephanie, the nineteen-year-old daughter of the head of the factory in Ghent where Kekulé went to buy gas for his lab. Stephanie, the female form of Stephen, from the Greek word for crown: "to encircle...that which surrounds."

A year after they married, Stephanie gave birth to a son. She died ten days later of puerperal fever, probably from the dirty hands of her doctor. She suffered. The infection spread from her uterus to her stomach, making it swell up, a monster pregnancy of infected fluid and pus. Convulsions swept her body; she hallucinated from fever; she felt so much pain she couldn't even stand for the sheet to touch her.

: :

The U.S. doctor Charles Meigs wrote a book in 1848 called *Females and Their Diseases.* Everyone considered him the top expert on childbirth. He described what people called childbed fever: "I have seen patients...in whose minds that pain appeared to incite the most unspeakable terror. I think I have seen women who appeared to be awe-struck with the dreadful force of their distress." For Meigs, it seemed that in their suffering, the women could see what other mortals could not—a god come to visit his wrath on her writhing body.

Some physicians tried to point out that many fewer women died when doctors washed their hands before delivering babies, especially if the doctor had just been touching a dead body. The medical establishment dismissed the idea that doctors could be contaminating mothers. Meigs protested that doctors could not possibly carry infection because they were gentlemen, and "a gentleman's hands are clean."

LOSS

On January 6, 1942, Janie Christman, mother of the PBY pilot, received a telegram at her home near Mt. Angel, Oregon, about forty miles from Portland:

THE NAVY DEPARTMENT DEEPLY REGRETS TO INFORM YOU THAT YOUR SON ENSIGN ELWYN LEWIS CHRISTMAN UNITED STATES NAVAL RESERVE IS MISSING RESULT AIR ENGAGEMENT WITH ENEMY IN THE PERFORMANCE OF HIS DUTY IN THE SERVICE OF HIS COUNTRY X THE DEPARTMENT APPRECIATES YOUR GREAT ANXIETY AND WILL FURNISH YOU FURTHER INFORMATION PROMPTLY WHEN RECEIVED X TO PREVENT POSSIBLE AID TO OUR ENEMIES PLEASE DO NOT DIVULGE THE NAME OF HIS SHIP OR STATION=

Christman managed to land his burning airplane in the ocean. He and the second and third pilots slid into the water through the navigation hatch. They pulled out the radioman, Robert Lee Pettit. The flesh had burned off his hands and face. As the four moved away from the flames, the plane broke in two and sank.

Of the seven-member crew, two had jumped before the plane hit water. They were nowhere in sight. One of them, mechanic Joseph Bangust, died after jumping. The seventh crew member, Andrew Waterman, also a mechanic, sank with the plane, killed by Japanese machine-gun fire.

The two mechanics operated machine guns in the waist of the plane. In most versions of the PBY flown by Christman's squadron, shooting at enemies required sliding open a metal hatch, sticking the gun out, and looking out of the hatch into the wind to aim. Later versions of the plane came with a new design: clear, hand-formed blisters for the gunner to look out of, probably made from polymethylmethacrylate, or Plexiglas, a brand-new plastic.

Once in the water, the survivors cut off their heavy wool flight suits and inflated their life jackets, which service members called the "Mae West" because it puffed up on their chests like large breasts. They had landed in the Sulu Archipelago some twenty miles from the nearest island. It was about 7 a.m. They began to swim, guiding Pettit, who could only float on his back.

By afternoon, despair had begun to set in. The sun blistered their skin and they burned with thirst. They decided that the copilot Bill Gough, the strongest swimmer, should go ahead and try to find help. Here, Christman's official report diverges from the story reported by the war correspondent Cecil Brown some weeks later. Though both record that Pettit could swim on his own by this time, the official report from Christman never mentions him again.

Brown writes that the three remaining (Christman, Pettit, and third pilot Don Lurvey) agreed each should go alone. "I'm in a hell of a mess," Pettit said. "You guys go on without me." Nonetheless, they swam within shouting distance of one another. Around sundown, Christman called for Pettit and got no answer; he had disappeared.

Christman and Lurvey swam through the night. Christman would fall asleep swimming and start to sink, then suck in water and splutter awake. He began to hallucinate. He dreamed he swam through oil; he wanted to sit and rest on a pipe. Lurvey kept him going, pointed toward land.

Around noon the next day, they spotted an outrigger canoe sailed by indigenous Moro men, who agreed to pick them up.

The men's tether to the U.S. military broke the moment their plane hit the water and sank. Without the shell of a machine to surround them and make them visible from above, they disappeared from the industrial world of war-making—four specks in a shining blank.

Christman's mother received a letter from the Navy dated January 9, 1942, offering sympathy for the "loss" of her son.

The word loss is Old English. In the modern sense, it means "failure to keep or hold what was in one's possession; failure to gain or win." The older word had a stronger force: it meant total dissolution, to break apart.

Christman never mentions his father in his letters, and the official correspondence from the military is addressed only to his mother, Janie. I have tried to imagine what it was like for her, receiving the news first that her son was missing and then, a few days later, that he had died half a world away from her small farming community in the Willamette Valley. I have visited the area several times. It's a short drive from my house. I don't know the exact spot of the Christman farm, but Christman's son Lance told me it was between the towns of Monitor and Mt. Angel, four miles apart, and seven miles from what is now Interstate 5 and an outlet mall.

The region is still rural, still dominated by farms, orchards, and plant nurseries, several of them large companies, prospering from the valley's mild climate and fertile soil, a legacy of ice-age flooding that deposited sediment from hundreds of miles away into the valley. I have driven the two-lane roads through this intensely cultivated landscape, the Cascade Mountains looming in the east, with the singular, snow-covered peak of Mt. Hood—which looks like a child's drawing of a triangular volcano—and, on clear days, the Coast Range visible to the west.

It turns out this valley has long been marked by mourning and loss, and by events far distant in space and time. Monitor, which is less a town than a crossroads, with a market, a bar, and a sheriff's station, was named for a Union ship in the Civil War. Settlers from Bavaria came to Mt. Angel in the 1880s, around the same time some Benedictine monks from Switzerland founded a monastery on the bluff above town. The monks named the town and the monastery after their Swiss home: Engelberg, Mount Angel. The town displays its Bavarian heritage with a Glockenspiel, several biergartens, and a large Oktoberfest in the fall. Even the self storage business has a Bavarian style sign. The stone Catholic church with its dark spire dominates the landscape, along with an old granary, now an antiques warehouse.

Generations of immigrants from Mexico and elsewhere in Latin America have also shaped the valley. Current day Mt. Angel, population 3,300, is nearly thirty percent Hispanic or Latinx, and some surrounding towns have majority Latinx populations, many working in agriculture. Since the 1960s, Russian Old Believers have come to the valley too, their presence marked by churches with onion domes. They are descendants of Russians who refused to adopt Orthodox church reforms in the 1600s. Many fled first to China, but the dislocations of World War II and its aftermath drove them out again. They found in the Willamette Valley a place to farm and practice their traditional beliefs, which drew others from around the world. Oregon now has the largest population of Old Believers in the country.

For thousands of years before any of this, the Kalapuya people lived in this valley. Place names reflect the distinct bands of Kalapuya who lived along 150 miles of river: Tualatin, Yamhill, Santiam, and others. A few decades before the Bavarians and the Benedictines arrived, the U.S. Government stripped the Kalapuya of their million-acre homeland and forced the remaining population, already ravaged by European diseases, onto reservations. Most went to the Grand Ronde Reservation in the Coast Range west of the valley, which mixed together more than thirty Indigenous tribes and bands. Today the Kalapuya are part of the Confederated Tribes of the Grand Ronde. Descendants carry on the stories and customs of their ancestors, and they founded the Komemma Cultural Protection Association to research and maintain Kalapuya culture.

Esther Stutzman is chair of the nonprofit and a Kalapuya storyteller. "I want to make sure to preserve the stories that were passed down to me from the women elders in my tribe," she told a reporter from *The Daily Emerald*, the student newspaper at the University of Oregon in Eugene, which sits on Kalapuya land.

Other leaders work to recover the tribe's archival history—fragmented, scattered, and distorted by a racist society. Stutzman carries knowledge recorded by the Kalapuya themselves, a precious inheritance. She has thirteen stories she performs in public. Others she does not speak, record, or write, except with her own people. In her tribal traditions, sickness and death can come to those who mistreat stories. They are a form of cultural wealth that the settler colonists could not take. Stutzman describes them as possessions, "private property," but their loss would carry the word's original force: dissolution of an ancient way of knowing.

THE ALBATROSS

The albatross filled with plastic suffered. Susan wanted to make sure I understood this.

Jen and I flew to San Francisco, which put about nine hundred pounds of carbon dioxide into the atmosphere. Susan invited us on a bright spring day into her office in the basement of the California Academy of Sciences, below the ground of Golden Gate Park. The basement serves as resting place for twenty-six million plant and animal specimens. The word comes from the Latin *specere*, to look at. This basement also houses artifacts (which are things made by people) and scientists carrying out the academy's mission: *to explore, explain, and protect the natural world*.

Explore, explain, protect. To boldly go. But Susan Middleton is not a scientist. She's a photographer. She had recently photographed the Academy's collection of specimens for a book called *Evidence of Evolution*. Mostly, though, she takes portraits of living plants and animals. In 2003, she and her colleague David Liittschwager traveled to Kure Atoll in the Northwestern Hawaiian Islands to document the birds and plants and sea creatures there.

Hawaiians call Kure "the elder," said Susan. It is the oldest of the Hawaiian Islands, moving like all the others slowly northwest on the Pacific plate, eventually to disappear beneath the water.

In fact, Kure the island is already gone. The atoll represents the submerged mountain's coral remainder, the skeletal base of countless translucent creatures eroded over time to fine sand.

The photographers almost didn't get there. At fourteen hundred miles northwest of Honolulu, Kure is, from the human point of view, the most remote atoll in the planet's most remote island chain.

For the albatross, Kure is one of the few places on Earth it comes to rest. The bird spends its life in flight, ranging over thousands of miles of water, sleeping and feeding in motion, landing only to breed on specks of ground surrounded by a liquid horizon.

In-spire. The albatross is filled with air: Tiny sacs pack the vault of its ribs, curving around its organs and extending through the narrow bones of its wings that span six feet, eight feet, eleven, the longest of any creature.

Flight. That's not the right word. The albatross glides, its shoulder and elbow joints locked into place. It avoids the muscle power of flapping; instead it snaps open its wings and dips into global wind flows to cross miles of liquid earth, sniffing out squid, crustaceans, and flying-fish eggs.

No one knew until recently where albatross really traveled. Now people tape transmitters inside the bird's upper back feathers, so that an antenna sticks out. It sends radio signals to orbiting satellites that use Doppler shifts to calculate the bird's position: jagged, erratic tracks following food across thousands of miles of open water. Eventually the transmitter battery dies, and the albatross path disappears from human sight.

: :

No people live on Kure. It appears in historical records mostly as a site where ships ran aground on the shallow reefs. The State of Hawai'i, which manages the atoll as a wildlife refuge, allows no visitors, except scientists and researchers. Susan and David got special permission to go. They caught a ride on a research boat searching out old shipwrecks. They intended to stay on the island three days. They stayed three weeks.

They arrived just before breeding season for the Laysan albatross, a goose-sized bird with a white body and a dark cape. It has a hooked orange beak it uses to spear prey, and pinkish feet. Dark feathers surround its eyes and shade onto its cheeks, making it look dramatic, serious. Hundreds of thousands descend on the Northwestern Hawaiian Islands in winter to mate and lay eggs.

They form a carpet so dense they look like a living snow.

Anyway, that's how a biologist described the breeding birds. Like Coleridge, I have never seen an albatross. Since it's an animal synonymous with motion, it doesn't keep well in captivity.

MAUVE

In 1857, a "mauve madness" infected Paris, then spread to London. The Empress Eugénie started it. She thought the color matched her eyes, and rich women all over the world copied what the French Empress put on her body.

María Eugenia Ignacia Augustina emerged from her mother on May 28, 1826, during an earthquake in Granada, Spain. Her father was a Spanish count who worshiped Napoleon Bonaparte. He joined the French army to fight against Spanish patriots, then went to Paris and served as one of its last defenders against Prussia and its allies in 1814. He was living in Spain under house arrest when Eugenia was born. Her mother, María Manuela Kirkpatrick, daughter of a Scottish merchant, took Eugenia and her sister, Paca, to Paris for school, to learn to be ladies.

In 1849, Louis-Napoleon, Bonaparte's nephew, first caught sight of Eugenia at the presidential palace. Immediately he wanted her. She refused, but he persisted. For years, Eugenia evaded him, even leaving France for a while. For her this was a matter of survival. Giving Louis-Napoleon access to her body would doom her to a kind of nonexistence: a mistress, unmarriageable, at the margins of French society.

In 1852, Louis-Napoleon held a referendum open to all the men of France on restoring the French Empire. It passed with overwhelming support, and on December 2 Napoleon III declared himself Emperor. He insisted Eugenia and her mother spend the Christmas holiday celebrating with him. For two weeks, Napoleon III tried relentlessly to seduce Eugenia. His attendants called it "the siege." The guests all watched and whispered. One wrote that a "frenzy of passion" had seized the Emperor, but Eugenia held him off. The gossipers said that when he begged for sex, she told him,"Yes, when I am Empress."

It was a bold demand, because Eugenia, though rich and a countess, had no royal blood. Napoleon had already proposed, through official channels, to marry the seventeen-year-old princess Adelaide, a niece of Queen Victoria. But on January 29, 1853, Napoleon wed the Spanish countess.

The people of France hated her. They considered her a foreigner, a gold digger. They booed and threw things at her during the wedding procession. But Eugenia transformed instantly into Eugénie, Empress of the French. Like her father, Eugenia believed in the myth of the first Napoleon: that only a strong ruler could save the revolution and create a more equal society. She thought Napoleon III had the best chance to realize this vision.

Unlike her father, she could not be a soldier. Instead, she used her body as a means of displaying the power and wealth of the Second Empire. She appeared in an ever-changing array of gowns, hand-constructed of tulle, silk, and velvet. She draped herself in the crown jewels. One guest wrote that she appeared armored in diamonds and "glittered like a sun-goddess."

The gowns formed Eugénie's persona and shaped the character of the Empire. When the cage crinoline appeared in 1856, she turned it into a worldwide craze. The steel hoops made it possible to build a massive, bell-shaped skirt, expanding the canvas on which to display her opulence—a word that comes from work and means wealth extracted from labor. Napoleon teased her about her cage. She declared she couldn't live without it. Like diamonds, it was armor.

The massive skirts that swirled around the Empress and her ladies appeared in a rainbow of colors produced in the same way they had been for hundreds of years: by laborious extraction from the bodies of plants and animals. Eugénie's mauve came from a few species of lichen, and from a new product of the global economy: bird shit, mined from rich deposits in Peru, first for fertilizer, then for the purple color that had once been extracted from snails.

When William Perkin tried to convince European dyers to use his new lab-produced color, he changed its name to mauve to associate it with French high fashion. But such a thing as a synthetic dye did not yet exist, and most fabric dyers didn't trust this new creation. Only one agreed to try it: Robert Pullar, who dyed silk for Queen Victoria. Perkin and Pullar worked together to refine the dying process for mass production. Queen Victoria wore Perkin's mauve to the wedding of her daughter.

By 1860, demand surged around the world for mauve, driven by the Empress and the Queen. Perkin's new factory at Greenford Green, west of London, was ready to supply the coal-tar dye, and he became a very rich man.

Soon after it went global, the fad for mauve was over. But now chemists understood the power of color. They created new shades, one after another. They gave the colors political titles to link them to the moment and make them popular: Magenta for the recent French victory at that town, and Bismarck brown, which came in a variety of shades based on macho Prussian moods: "content," "enraged," "ill," and "icy."

Perkin's teacher, August von Hofmann, at first dismissed the new dye as frivolous; he saw no value in it. He underestimated desire, the frenzy of passion around the world for the glittering Empress in her cage of color. Perkin had found the first product with global demand to be made from coal tar. His discovery opened the way for drugs, fertilizers, and plastic—nearly every product now created for the global economy. Lust launched industrial chemistry.

THE PHOTOGRAPH

I die therefore I am

—Judith Goldman

The photographers Susan Middleton and David Liittschwager returned to Kure in May 2004. They found albatross chicks sitting in nest bowls on every inch of ground, waiting for their parents to come back with food from the ocean. The chicks had advanced to that odd, in-between state called adolescence, fluffy grey down erupting in awkward bits from sleek, adult feathers.

The photographers set up shop in an old Coast Guard shed, a small square of shade in the glaring white. Albatross evolved without mammalian predators, and the chicks viewed the humans as objects of curiosity, not a threat. They behaved like toddlers, Susan wrote, putting everything in their mouths, tugging on tent lines, biting shoes, and snatching bits of clothing left out to dry.

One chick nested just outside the shed. "We passed it every day when we came out to photograph," said Susan. She uses "it." As far as she knows no one ever determined this bird's sex. "I named it Shed Bird," she told me. "We said hello and talked to it. You do that kind of thing when you're out there for eight weeks. We watched it get bigger and bigger."

Shed Bird started to spread its dark wings, preparing for a life in air. "When the winds come in, the fledglings start to sense it," said Susan. "You can tell they are realizing they have wings and what they might be for. They jump and hover for just a few moments. When we saw that, we knew pretty soon Shed Bird would be gone."

One day David found the bird panting in the hot sun. It fell over and didn't have the strength to get up. He moved the bird into the shade, sprinkled it with water, and cooled it with a fan. Shed Bird seemed to revive. But the next day, it had died.

The manager of the Kure Wildlife Sanctuary decided to cut Shed Bird open. Susan watched her make the incision. It revealed a stomach stretched tight and perforated in two places. "Then she took the knife and actually opened the stomach. It was completely impacted, full of plastic: disposable cigarette lighters, several bottle caps, an aerosol pump top, shotgun shells, broken clothes pins, a little toy like a spinning top."

Susan wanted to know exactly what killed Shed Bird. She put on gloves and began to pick out every piece of debris. The shards got so small she needed tweezers. Three hours, flies swarming, the stench of rotting.

Susan removed half a pound of junk from the albatross chick, mostly plastic. "That bird was sturdy to have survived as long as it did," she told me. We spoke over the phone, I at home in Portland, and she in San Francisco. I recorded her voice on a tape that I could then play back, typing the words as she spoke. Somehow this tape (old cassette technology, not digital) got tangled in its plastic case and broke, so now I can't go back and hear her voice. This is what I typed:

A. Why did you pull the plastic out of the bird?

S. It was pretty emotionally driven, but it was also task driven. The primary impulse was curiosity. I wanted to know what was in there. I could see when she opened the stomach, I could see that it was mostly plastic, but I thought, "I want to see every single thing that was in the stomach, everything. Every last thing."

A. Why?

S. I wanted it to be seen. That bird was stuffed with plastic, which had led to dehydration, malnutrition, and ultimately starvation. Obviously, it suffered. It couldn't pass its stomach contents because it was so severely impacted, it couldn't regurgitate, and it couldn't accept food. That must have been a horrific experience for that animal. All the time it was spreading its wings, testing the wind, it was suffering; those sharp bits of plastic had perforated its stomach.

[Silence]

S. We got to know this bird, we saw it every day for two months. I felt this responsibility to document it so that people would see it. That, I thought, was the least we could do for it.

Susan spread all the pieces out, more than five hundred. Then she took a photograph.

GYRE

Here is what I learned about Kamilo Point, Hawai'i, from Dr. Curtis Ebbesmeyer, an oceanographer who is a friend of Noni's. This small point of land sits at some kind of exit point for a vast gyre that circles the North Pacific. Winds and the spinning planet set in motion currents that travel around Earth's oceans. Five gyres circle each main ocean basin: North and South Atlantic, North and South Pacific, and Indian. Ebbesmeyer calls them the planet's greatest features after the continents and the ocean itself. "The gyres form continuous loops, like a snake biting its tail, but they are composed of distinct currents, like the vertebrae."

The spinning spines move more water than all the planet's rivers. Ancient sailors traveled them like highways. Nineteenth century adventurers used drifting bottles to plot their shapes. Twentieth century lab tests and satellites confirmed these maps. They drive climate and commerce and culture. They stir the oceans, bathing the northern coasts with warmth, carrying salt, people, and fish, and the endless multiform drifters that make the ocean a living broth. These are the tracks the Laysan albatross follows to sniff out its catch.

Until recently, only oceanographers talked about gyres. Now the word "gyre" seems to be on everyone's lips because of the discovery that plastic accumulates in their still centers. Researchers and volunteers sail out to try and find how much plastic might be out there; inventors look for ways to scoop it back out of the water.

Kamilo has always been a delivery point for things from the ocean. The first Hawaiians kept watch there for giant pine and fir trees carried from the West Coast of North America, from Oregon, where we live in a house built sixty years ago from old-growth Douglas fir. The Hawaiians used the timber not for houses, but canoes, one hundred feet long. This gave them military power.

Ebbesmeyer calls the trees that washed ashore "the keys to wealth and war."

Today, the same currents pick up debris from every nation that touches this water. A 2015 study in *Science* concluded that the top six countries releasing waste into the ocean are China, Indonesia, the Philippines, Vietnam, Sri Lanka, and Thailand. The trash from those countries often originates across the world, exported by places like the United States, Canada, and Britain. Before China stopped taking U.S. plastic for recycling in 2018, the U.S. sent four thousand shipping containers of waste every day to its shores. Now Indonesia absorbs more of this flow of trash, which lines roadsides and rivers in Java. Some of it washes up at Kamilo, twenty-four hundred miles from the closest continent. Some of it stays out in the gyre's spinning center.

From our trip to Kamilo, we kept very little. Only what we could squeeze into the Jeep and then into luggage later. I kept a faded fishing float with Korean lettering, a pale blue detergent jug with Japanese characters, the two XYLICUBE gum containers, the Astroturf scrap, a melted glob of whitish waxy plastic shaped like a roosting dove, the Epson cartridge, a feathered plastic wing torn from its toy bird, and a liter bottle with a turquoise cap that Jen filled with a confetti of plastic bits gathered from around her as she sat on the sand. As I turn it around in my hands, a few recognizable objects appear: a white toothpaste cap, a scrap of blue fishing net, a clear round ball from a deodorant stick.

At Kamilo, Ron drove his Unimog over lava right to the edge of the water, hooked a massive heap of fishing net, line, and rope to the back, and dragged it, engine roaring, up the beach above the tide line. Noni and Ron bought the truck for this purpose. Fishing gear—lost or tossed off boats—tangles in giant snarls that roll along with the currents. People call these "ghost nets" because they continue to fish, snaring turtles, birds, whatever drifts too close.

The Hawai'i Wildlife Fund organizes regular beach cleanups here. Its website says volunteers removed one hundred tons of plastic over the past four years. Noni confirms that things are much better now. When they first came down twenty years ago, they found a wall of debris piled several feet deep.

But no one collects the fragments. No one could. It would be like gathering the sand off the beach. Bright bits drifted in crevices and pooled in depressions in the lava. I lay on my side near a small, protected shelf and watched the waves wash in, bright shards swirling in the water. Some stayed behind, shining on the black lava; some tumbled back out with the waves.

WHITE WHALE

The trip to Kamilo paralyzed me for awhile. I thought that to write this book I would need to travel to every junk beach on the planet: places where spinning gyres, currents, and geography coincide, and the ocean vomits up some portion of its vast cargo.

Then I would need a submersible to take me down three thousand feet into sea canyons in the Mediterranean, where plastic bottles pile up in the sediment like strange cylindrical fossils. Then across the seabed and downward, darker, colder, eight thousand feet below ice floes in the Greenland Sea, my arc of light reflecting off a clear plastic bag floating past in the black.

Then smaller and smaller still until I become microscopic, small enough to slide inside the gut of a lugworm, *Arenicola marina*, burrowed beneath beige sand at Dovercourt Beach on England's east coast. A powerful suck from the worm's mouth swirls me inside it, slick and dark, along with sand grains and tiny plastic particles.

The worm's muscled gut walls contract, stripping from the miniscule bits whatever its cells recognize as food, whatever is organic—bacteria, algae, protozoa, and the laboratory chemicals that saturate the particles—polychlorinated biphenyls, dichlorodiphenyldichloroethylene, nonylphenol, phenanthrene, triclosan.

Working the plastic through its guts makes the worm tired and sick; its walls contract less and less. It rests. A bar-tailed godwit catches sight of the worm's dark tip, snatches with its narrow beak, swallows in two quick jerks. The worm melds into bird cells. As for me, I am ejected, only imaginary, cast back on the beach in a tube of sandy worm excrement.

: :

I lay on the couch on a Friday, the 13th of January, thinking such thoughts. I felt incapable of writing this book, of expanding to the necessary scope. Of acquiring the time and money and breadth and depth of—what? I didn't even know what I was missing.

The Autobiography of Plastic. A title I announced to my friends on a whim and kept repeating because people liked it, enough to give me a few thousand dollars in grant money. Enough to interest a book agent, who asked for chapter summaries. But I had no idea what it meant. I could not give her a map to what I was trying to understand myself.

January in Portland means evening all day. It had rained, and it was going to rain. But for a moment the sun burned almost all the way through the cloud layer and turned the sky silver. I decided to stop thinking, get off the couch, and take the dog for a walk. I opened the front door and there in the dirt sat a white plastic ring, gleaming. I picked it up. It was deeply worn and weathered. Hard to tell where it came from, the mouth of a pill bottle maybe. It had teeth marks in it, definite marks from some animal's mouth.

It struck me: I am not worm shit. Imagining piloting around, the watcher peering out from her safe suit of self—that is wish fulfillment, a fantasy. I am inside the worm still, the worm inside me, sloshing molecules back and forth. A house forms its own coast, a body does—skin, blood, gills, lungs—awash in the currents and whatever they bring, seeping through cells respiring, tidal.

The teeth that chewed on this plastic probably belong to our dog, Quincy. He chewed on it, and swallowed some tiny plastic particles, and some drifted down into dirt, where in about one month, we will poke a few holes and tuck in pale peas that will grow into plants, vining themselves out of sun, water, air, and nutrients pulled through roots from the soil.

In May, I'll walk out and pinch off a sweet green pod, chew it up with my teeth, and swallow. Quincy the dog will do the same;

he learned it from me, tugging pods with his mouth. We always plant too many, so I will fill bowls with peas and share them with neighbors, who will admire the sweet crunch in their teeth. In there, along with whatever molecules make a pea, there might be a few broken free from the plastic bits, and whatever else has washed this coast in its sixty years as suburban tract: particles of soot from car exhaust, bits of mercury fallen with rain drops, asbestos slivers from the house shingles.

Here neighbor, let me feed you, let me feed you, dog, let me feed you, hungry body.

: :

I put the ring in my pocket and walked out the gate. I felt a little whisper of that vertigo. Our neighborhood that had been familiar, kind of weedy and worn but tidy, turned up a new skin, studded with plastic: street, sidewalk, lawn, gutter. I picked up each piece and put it in my pocket. Then I couldn't stop. Every day, on every dog walk, with disgust, boredom, sometimes delight, I picked it all up. I began to make lists, recording each piece and taking a photograph. An ocean flooding my own shores, I absorbed every bit. I stored it all in plastic garbage bags on the back porch.

In October the car part appeared. "That's your white whale," said Jen when I dragged it in. I admit: I have never read *Moby Dick*. I searched "white whale," just to see what would turn up. The definition from urbandictionary.com:

> Something you obsess over to the point that it nearly or completely destroys you. An obsession that becomes your ultimate goal in life; one that your life now completely encircles and that defines you.

Nearly or completely. What nearly or completely destroys you.

DISAPPEARANCE

Since his arrival in the U.S. in 1935, Stan Ulam had felt taken aback by the apoliticism of American academics. He remembered running into the logician Willard Quine on the steps of Harvard's Widener Library the morning that Franklin Delano Roosevelt won the presidency. Ulam inquired what the philosopher thought of the election. "Who is President now?" Quine asked him.

When war broke out, Ulam bought every edition of all the papers, hoping for news of Lwów. He came across a picture of his brother Adam in *The Boston Globe*, surrounded by other Brown freshman, with the photo caption: "Wonders whether his home was bombed."

Denmark, Norway, the Netherlands, Belgium, and then France fell to Hitler in the spring of 1940. "Despair gripped all the European émigrés on this side of the ocean," writes Ulam. The U.S. persisted in its isolationism; Americans believed their oceans would protect them. For the European Jews, it seemed even three thousand miles of water might not be enough to douse those flames.

Through friends at Harvard, Ulam got a job teaching at the University of Wisconsin in Madison. For an upper-class European intellectual, the U.S. Midwest seemed like Siberia. He felt restless, agitated. He wanted to do something for the war.

As soon as he could, Ulam earned his U.S. citizenship, then volunteered for the Air Force, which had advertised on campus. At each step he encountered others displaced. He took his citizenship test from a Jewish man whose parents had come from Ukraine. He received his Air Force physical from Japanese-American medics forcibly "relocated" to Wisconsin from the West Coast. Ulam had a blood test for the physical, and he makes a joke about "losing blood to the Japanese" in defense of his new country.

Blood, from the root

"to swell, gush, spurt. That which

bursts out." See

bloma as bloom, "to flower." Body

as fact flowered

blood and skin flowered

hair bloomed bone. Here

the hyphen a

Japanese

-American, a Jewish-

American. Fact. A failure

to disappear—

The U.S. military would not accept Ulam. He lacked binocular vision: His right eye was myopic, something he had always kept hidden.

The page in Ulam's memoir that recounts this is followed by a reproduction of a pencil sketch of him done by his cousin, the artist Zygmund Menkès, in 1938. The sketch also appears on the cover of the book.

How did I not consider this drawing when I was thinking about Ulam's eyes? It sizzles with kinetic energy. Quick slashes and scribbles define Ulam's cheeks, brow, and hair; a squiggle even floats above his head, as if he were giving off smoke.

His eyes form the dark center, concentrated pools of scribble beneath two slash brows. The pupil of the right eye (the myopic one) sits farther up than the left, partly obscured by the lid, leaving that eye mostly white, a blank.

Ulam's flawed eyesight left him stuck in Wisconsin, teaching math to Navy recruits as his wartime contribution. He burned for something more. He noticed that people had started to disappear from the school—physics professors and students. He wrote his friend, the Hungarian mathematician Johnny von Neumann, who had been spending a lot of time in Washington, D.C. He mentioned the disappearances of his colleagues. He wondered about possible wartime work.

In the fall of 1943, Von Neumann had Ulam come to Chicago and meet him between trains at Union Station. His friend arrived at the tracks accompanied by two armed guards. This impressed Ulam. Von Neumann described a secret project taking place somewhere in the American Southwest.

A few months later Ulam received his invitation to P.O. Box 1663, the code name for Los Alamos. The job involved "an unidentified project that was doing important work, the physics having something to do with the interior of stars." He had engineered his own disappearance.

Kekulé felt he was drowning in grief after Stephanie's death. He wrote his former student Adolf Baeyer: "I'm sick...From time to time it gets a little better—ebb and flow—and then I can work at least a little...I will make the most desperate attempts not to sink."

By January 1865, Kekulé apparently still felt unable to speak. His friend, the chemist Adolphe Wurtz, presented his discovery about the structure of benzene to the Chemical Society in Paris. Kekulé's vision of the whirling snake held the key to a new world of molecules.

Chemists at the time debated the existence of atoms; they had no inkling of electrons. They knew particles came together into molecules, and that the shapes of molecules mattered to the resulting material, but how and why fueled fierce controversy.

Kekulé had been led by a previous vision of dancing figures to the idea of valence, that particles of certain types always bond to the same number of other particles: hydrogen to one, oxygen to two, carbon to four. He dealt with benzene by imagining alternating single and double bonds between the carbons, but that left two bonds extra. Inspired by his vision of the snake, he connected them to form a circle.

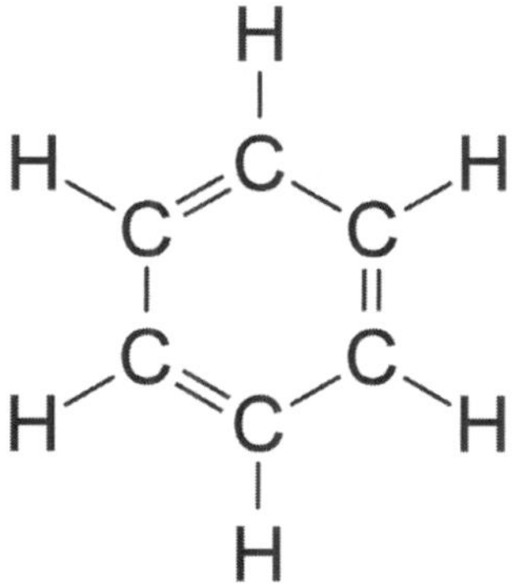

Like all organic molecules, benzene depends on the special properties of carbon. It links easily to itself and other atoms, but once connected, its bonds remain relatively stable. This makes carbon ideal for forming and reforming the complex molecules living beings require, and for building new molecules in the laboratory.

Kekulé's revelation gave chemists the power to predict exactly how the molecule would act. Atoms left the world of ideas, of philosophic speculation, and became a set of parts to be manipulated into useful structures. Chemists could finally "see" what they were doing.

But Kekulé's ring did not explain all the mysteries of benzene. It remained more stable than its shape could account for. He missed an important feature of his vision: The snake was alive, it whirled in constant motion.

In the 1930s, the chemist Linus Pauling proposed a new idea: The electrons that form the bonds never appear in a single place but shift constantly among the carbons, causing the molecule to oscillate. The ring shape in constant motion gives benzene its stability. It never takes a single form, flashing on and off like a ghost among all of its potential states. Pauling called this "resonance," from the Latin word for echo.

: :

The benzene ring vibrates inside the cells of every living being: plant, animal, human, and their fossilized and liquefied remains—coal and oil. Torn and scraped out of ground, coal fueled the industrial revolution's smoking factories. Benzene split from coal fueled the labs, where chemists created molecules that had never before existed.

They chemically bolted these creations together to see what they could make. Out of their labs came dyes, drugs, pesticides, explosives, and something else, a new material. It could be melted and formed into whatever shape a mold could take. Heat and pressure transformed it into an object, or millions of identical objects, impervious to flame, corrosion, electricity, water, decay, or other destructive force. Its first trademark symbol: infinity.

GUSH

> *Matter is pitiful; form is terrible.*
>
> —Lisa Robertson

The albatross hardly appears in the poem by Coleridge. He names the bird seven times in six hundred lines: when it first becomes visible, flying through fog, the only living thing the sailors see in the land of ice; when it follows the ship for nine days eating human scraps; and when the mariner shoots it with his cross-bow.

At no time does Coleridge describe the bird or give any sense of its physical presence—not even in its fourth mention when the others hang the albatross around the sailor's neck. The dead bird, lying against his heart, must have been an awkward weight; it must have stunk as it decayed, but Coleridge remains silent on these points.

It is the sea serpents that burst to life inside the poem. The mariner recoiled from them at first, monstrous, slimy things. After he is the only living being left on the boat, he looks closely:

> Beyond the shadow of the ship,
>
> I watched the water snakes:
>
> They moved in tracks of shining white,
>
> And when they reared, the elfish light
>
> Fell off in hoary flakes.
>
> Within the shadow of the ship
>
> I watched their rich attire:

> Blue, glossy green, and velvet black,
>
> They coiled and swam; and every track
>
> Was a flash of golden fire.

There follows a gush: *a copious or sudden emission of fluid; a rush (of water, blood, tears).*

It is not the poem's first. The first gush aborts. It dries up. The sailor is surrounded by bodies on a rotting ship on a rotting sea:

> I looked to heaven, and tried to pray;
>
> But or ever a prayer had gusht,
>
> A wicked whisper came, and made
>
> My heart as dry as dust.

: :

Heidegger also has a gush. In his meditation on the clay jug, he writes that its character consists in holding in and pouring out. The pouring is a giving, he says, sometimes for people and sometimes in honor of a god:

> "Gush"...is the Greek *cheein,* the Indoeuropean *ghu.* It means to offer in sacrifice. To pour a gush, when it is achieved in its essence, thought through with sufficient generosity, and genuinely uttered, is to donate, to offer in sacrifice, and hence to give.

Heidegger uses the German word *giessen*, to pour, which his translator renders "gush." The English word "gush" has a murkier origin. It appears in Middle English with no clear antecedent in Old English or any other Germanic language. The Oxford English Dictionary presumes it is onomatopoetic and cites as its earliest source the poem *Morte Arthure,* written by an unknown author around 1400. The reference is a violent one, from Arthur's fight with the giant:

Bothe þe guttez and the gorre guschez owte at ones.

Other sources say the word appeared earlier. According to vocabulary.com: "Gush comes from the twelfth century English word *gosshien*, originally 'make noises in the stomach,' and later 'pour out.'"

: :

Albatross chicks have to grow very fast. They need to mature enough in just a few months to fledge and spend several years in flight without touching land, hunting their own food. The digestive coil inside torments them without ceasing. The parents hear it also, this gush; it pushes them out thousands of miles across the ocean and reels them back. The adults take turns feeding, making constant calculations about how long to leave the nest, how far to fly and in what direction to find enough food to fill the ravenous hole.

: :

As I read about the albatross, I also was reading *The Odyssey*, in its new translation by Emily Wilson, the first woman to publish a major translation of the ancient poem in English. I am struck by the fact that hunger also haunts Odysseus throughout his journey. It brings death to his men who, starving after being marooned for a month, kill and eat the forbidden cattle of the sun god. The flayed hides begin to crawl, and the meat moos on its skewers. Still the men eat, feasting for six days. Once they are back at sea, Zeus hurls a thunderbolt and kills them all for their transgression.

Hunger propels Odysseus to humble himself and beg; its animal need makes him both dangerous and vulnerable. When he washes up on the island of the Phaeacians, he approaches the princess and her slaves:

> Just as a mountain lion trusts its strength,
>
> and beaten by the rain and wind, its eyes
>
> burn bright as it attacks the cows or sheep,
>
> or wild deer, and hunger drives it on
>
> to try the sturdy pens of sheep—so need
>
> impelled Odysseus to come upon
>
> the girls with pretty hair, though he was naked.

Hunger is cruel; it drives the body toward survival. Odysseus tells the Phaeacian king, Alcinous:

> The belly is just like a whining dog:
>
> it begs and forces one to notice it
>
> despite exhaustion or the depths of sorrow.
>
> My heart is full of sorrow, but my stomach
>
> is always telling me to eat and drink.

: :

Albatross are far older than humans on the Earth, descendants of the dinosaurs. They evolved over thirty-five million years into finely honed gliding beings with bills curved to hook prey. The Laysan albatross mostly spears squid, but it also collects floating chunks of pumice and wood with fatty, nutritious flying fish eggs attached. Bits of plastic gather in the currents along with stone and wood, and the albatross picks them up.

Like other birds, the albatross regurgitates what it can't digest. Before it fledges, the chick sheds weight by coughing up a cigar-shaped lump called a bolus filled with squid beaks, stones and, increasingly, plastic. But plastic doesn't wear smooth in the waves. It retains its shape or breaks into shards, and sometimes the bird can't get it out.

: :

A new image of an albatross has etched itself inside my brain. This one comes from the scientist Carl Safina. He traveled to Midway, an atoll in the Northwestern Hawaiian Islands about fifty miles southeast of Kure. Midway is home to the world's largest Laysan albatross colony: 600,000 breeding pairs.

In his book *Eye of the Albatross*, Safina describes watching an adult glide in from the ocean, pick out its chick among the crying and begging thousands, and open its bill to deliver food into the thrusting, frantic mouth. "The adult hunches forward, neck stretching, retching," he writes. It delivers several chunks of "semi-liquefied squid and purplish fish eggs."

The adult continues to retch and the chick to beg for more, but nothing comes. Then Safina sees the tip of a green toothbrush emerge from the bird's throat. The bird tries several times to get the toothbrush out with no success. It gives up and wanders off.

An albatross can live for sixty years. A plastic toothbrush can last—no one knows how long. Five hundred years? A thousand? How long can an albatross live with a green toothbrush stuck in its gullet?

: :

> O happy living things! no tongue
>
> Their beauty might declare:
>
> A spring of love gushed from my heart,
>
> And I blessèd them unaware:
>
> Sure my kind saint took pity on me,
>
> And I blessed them unaware.

Awed by the beauty of the sea serpents, the mariner in Coleridge's poem prays, and the albatross drops from his neck and disappears into the ocean. The bird is not a burden. It never was in the poem; it never had that physical heft. The albatross around the neck is a mark of guilt.

The mark remains, even after the body of the bird is gone. The sailor is compelled to spend the rest of his life telling others what he has seen.

DESIRE

In 1988, *The New York Times* published the first-ever image of a benzene molecule. It looked exactly as Kekulé had envisioned it one hundred and twenty-five years before.

Kekulé may have made up the story about his vision. He told it only once for the public, late in his life, nearly thirty years after the event. Some suspect he related the story of the whirling snake to solidify his primacy in its discovery.

But Kekulé's story persists as one of chemistry's founding myths, and the image of benzene confirmed his prescience. Scientists at IBM created the picture that appeared in the *Times* with a scanning tunneling microscope. The instrument uses an atom-sized probe to track variations in electrical current between it and the surface being mapped.

The resulting figure looked startlingly like Kekulé's daydream and the simple hexagonal line drawings used ever since to represent benzene. Such an image "gives the feeling that one could simply reach out and touch the atoms," said the chemist David Goodsell.

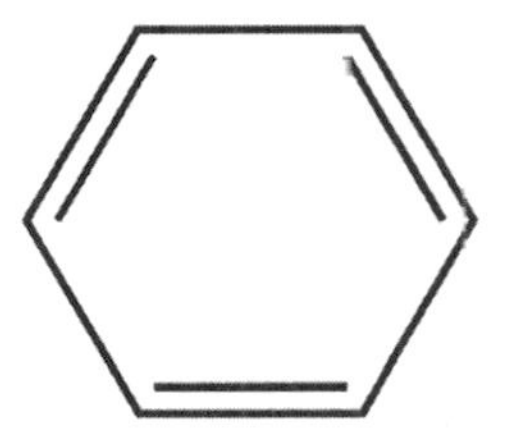

But no one has ever seen or felt an individual atom, and no one ever will. Microscopes do not passively record pictures of molecules. The process requires a complex set of interactions among scientists, technicians, machines, and materials. The result is a series of numbers "unreadable to most people, even scientists," writes the chemist Tami Spector. Technicians use computers to transform the numbers into an image, a sort of topographical map of electrical signals.

Researchers, scientific journal editors, and the press blank out these factors in order to perform the fiction of sight. They present the figure as if it were an actual picture of atoms and molecules. Scanning microscopes yield numeric data that people transform "into images disguised as photographs," Spector writes.

Why? Desire. To reveal, to know, a word that is very old, and which Western culture since Plato has linked with sight. In philosophy, this is called ocularcentrism. To look with one's own eyes. To see for oneself. To see and believe, which comes from the root for "desire."

"The experience was so sublime that it sent chills through my body," writes the physicist Karen Barad, on first seeing individual carbon atoms imaged by a scanning tunneling microscope.

The same hunger to know consumed Kekulé. He thought of molecules constantly. He dreamed about them. He pictured molecules and then built them using sticks and balls made of wood and metal.

The diagrams he and others pioneered remain the common currency for conveying chemical information. But Spector notes they are really, as their origins in wood imply, stick figures for something far more complex: "They communicate a sense of the molecule's spatial geometry... [but] reduce experimental evidence, stripping it bare of noise, impurities...presenting instead an idealized abstraction of a single, motionless molecule."

I wrote Dr. Spector an email. Her name conjures an image in its most ephemeral sense, a ghost. She told me that what scientists want most, even more than pictures of atoms and molecules, is to capture them in the act of transforming. It is the instant of change they most long for.

In spring 2013, the Lawrence Berkeley National Lab issued a press release headlined, "A chemical reaction caught in the act." It claimed scientists had captured the first-ever high-resolution images of a molecule—benzene rings—breaking and reforming chemical bonds.

The point of the research was to try and engineer graphene, a single layer of carbon atoms bonded in hexagons. That failed, but the news was not the research, it was the imagery. The press release quotes lead scientist, Felix Fischer: "We weren't thinking about making beautiful images; the reactions themselves were the goal." But, he adds, they needed "to really see what was happening at the single-atom level."

To really see. But Dr. Fischer uses a metaphor of blindness, comparing the atomic-force microscope moving over the surface to reading Braille. "The resulting images," the press release reports, "bore a startling resemblance to diagrams from a textbook or on the blackboard, used to teach chemistry, except here no imagination is required."

No imagination required. Science has vanquished the blank, the gap between mind and world, delivered the un-seeable to the human eyeball on a surface of silver. Says Dr. Fischer: "What you see is what you have," which means "possess," and comes from the root word for *grasp*.

Of course, the reaction itself can't be captured in a still picture. What the images show instead is the "before" and "after" bridged by an arrow. All of the meaning hangs on that black connector. Carry us there, O strike us, arrow, arrow. *Lust of the flesshe, lust of the eyes.* "All images are after."

DESIRE

The car part says HONDA in large letters, and it contains other markings, mostly incomprehensible to me. One can ask the Internet anything, and it almost always supplies a certain kind of answer, but it had nothing coherent to say about this set of numbers and letters.

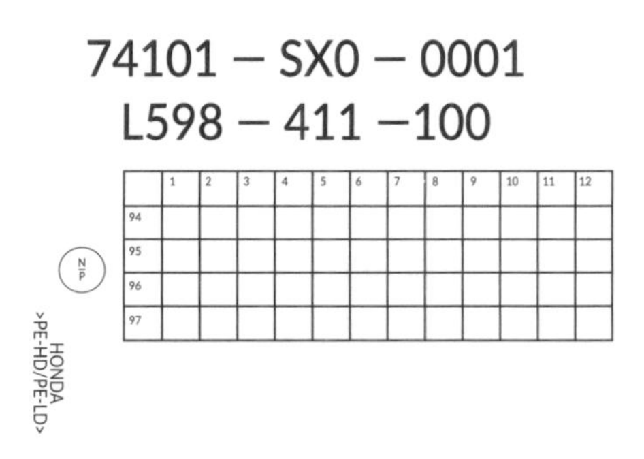

I asked around a little bit. I emailed this code to people who know something about plastic, and something about auto manufacturing. No one could tell me much, but a chemist I know, who is an expert on plastics, told me that the PE-HD/PE-LD means the car part is made of polyethylene, a mixture of high- and low-densities. He suggested I try talking to someone at Honda.

The nearest Honda manufacturing plant is probably Greensburg, Indiana, two thousand miles away. But I found something called American Honda Motor Company on Northeast Sandy Boulevard here in Portland. The listing included two Portland phone numbers. After getting no answer from the first, I tried the second. This time a computer picked up with a woman's voice saying, "You have reached American Honda Motors," before giving another number I could call "in case of emergency."

I hung up. What kind of emergency could someone have that would require calling Honda Motor Company? I decided to try the number. It had a Los Angeles area code. Someone answered after only two rings. His voice was so fast and clipped that at first I thought he too was a computer: "Honda corporate security. This is Keith. Are you calling to report an emergency?"

My first instinct was to hang up, but then I remembered that sometimes calling 911 and hanging up can trigger the cops to come anyway. I imagined some kind of corporate security force (guys in suits?) showing up at my door. "Oh," I said, sounding confused, "I guess I have the wrong number." "OK," said Keith, "that's no problem." He said it in a tone that sounded vaguely threatening, like it's not a problem *this* time, or, it's not a problem, but if he wanted to, he could make it one.

I abandoned the direct approach and turned to spying through the Internet. Honda Motor Company on Sandy wasn't marked on Google Maps—Street View showed a large, square facility, larger than any other building nearby, with a flat white roof punctured by small skylights, many shadowed by what looked like white crosses, just beneath the glass. The building bordered a housing development cushioned by a thin line of trees. A meticulously landscaped lawn faced Sandy. Next door was a strip mall with Teeny Foods, Skinny Mint, Strategic Pharmaceutical Solutions, and Portland Specialty Baking. Across the street was *Iglesia Nuevos Comienzos*.

It occurred to me that Google Maps is like starlight: It felt like looking down at what was real right then, but it could just have been an image of something no longer there.

On Monday, I tried the Portland number again for American Honda Motor Company. A woman whose name I didn't catch picked up. I told her I found a large Honda part in my front yard. I explained that I was writing a book about plastic and that I'd like to trace the origin of this car part. She listened in silence, then transferred me to the parts department. Selena picked up. I told her my story. She suggested I call Honda corporate headquarters.

"What do you make there?" I asked. "We don't make anything here," she told me. "We're a parts supplier for dealers."

I decided to be a little persistent, "Oh, so you don't think that if I brought in the part (or read to someone there what was on it), they could tell me anything about it?" She asked me to tell her what it said. She sat in silence while I read off the long strings of numbers. "The first one is a part number," she said. "The rest I have no idea. These come from Japan."

It was a pronouncement that seemed to seal the mystery. "Do you think anyone else there might know?" "Let me check with my manager." She was gone for a while. I could imagine the conversation taking place in some office nearby. She came back. "My manager says you need to talk to Honda Corporate. They can answer all of your questions and give you everything you're looking for."

But I knew that Honda Corporate could not give me everything I was looking for. I brought this dirty plastic car part, this piece of garbage, into our house. It sat in the living room for a long time, until Jen asked how long I planned to keep a car part in our living room. I moved it into the bedroom, on the floor near my side of the bed. It made a flat "clap" against the floorboards when I tripped over it, which I did almost every night.

I wanted it to speak to me. I wanted it to tell me something about how to live. How to live now, on this planet, in this real life, as a member of the human species. I wanted it to tell me what to do. What to do about being alive in a dying world. Where being alive is monstrous. Where the terms of every breath seem to be death.

I wait.

For what—

a sign. For an albatross to wake from my chest and take flight.

But it remains here with me. Me and this car part, its dirty carapace curled around me. There is no gap between us, no other "out there" to access, by microscope or imaginative vision. Here we are. Together. And the industrial chemicals we share, the resonating molecules of our bodies.

RESURRECTION

Elwyn Christman did not die as the U.S. military reported to his mother. But he was barely alive after his plane crashed and he had spent thirty hours in the water.

He and his copilot Don Lurvey's survival hinged on their encounter with a boatful of indigenous Moro men in the Philippine archipelago. The Moro regarded them with well-justified suspicion. The Japanese had reportedly entered Jolo harbor on Christmas Day flying an American flag. When residents crowded onto the dock to meet the ships, they opened fire.

Christman told the war correspondent Cecil Brown that it took an hour, the Moro men keeping a safe distance in their boat as he and Lurvey treaded water, to persuade the men they were allies, from the U.S. Navy. Christman used the few words of Tagalog he picked up during his two years of Naval Reserve service in the Philippines. He finally succeeded by pulling off his lifebelt and showing them the stamp: U.S.N.

Even convinced the men were American, the Moros still had reason for suspicion. After the U.S took the Philippines from Spain, soldiers fresh off the Indian Wars spent the first decade of the twentieth century marching across these islands "pacifying" the indigenous people by massacring them. It is unlikely that Christman and the others had any knowledge of this history. The Moro men would have been acutely aware. Similar encounters had played out countless times here—they almost always ended in gunfire.

I found no record of this meeting from the Moro point of view. To the men of the VP-101 squadron the Moro people seemed utterly inscrutable: at times full of generosity, at times terrifying, with their filed teeth and shining kris daggers. "We were never sure what the Moros intended to do with us," wrote Lieutenant Jack

Dawley, pilot of a second plane in the squadron that crash-landed in the islands.

The surviving crew members of the two planes managed to reunite on the island of Siasi. Over the next few weeks, they depended on the Moro people completely. Though they encountered various Filipino officials, Chinese merchants, and a few Americans who helped them, only the Moro men knew the tides and currents and could safely thread their vinta canoes through the archipelago's hundreds of islands.

The Moro sailors did not want to travel far from their villages in the midst of a Japanese invasion. Several of the VP-101 men had suffered burns and wounds and could not travel well anyway. One, Evern McLawhorn, had seven bullet wounds in his arms and legs, and a metal splinter embedded in one eye. They crossed the archipelago slowly from island to island, pausing to recuperate and negotiate for new boats and Moro sailors at each stop. Lieutenant Dawley submitted a fourteen page, typed "Supplement" to his official report describing their journey.

For the VP-101 men, stripped of their uniforms, weapons, airplanes, and radios—all the technologies that tied them to globe-spanning powers—time began to dilate, identities to shift. They wore "Moro" clothing. They slept in bamboo nipa huts. The hot mornings they spent drifting in boats, afternoons when the winds picked up sailing, and many nights rowing with tides and currents. They passed New Year's Eve on the water but had no way to know when midnight struck; they marked it as the moment the moon rose up over the mast.

Christman and his squadron mates never ceased in their efforts to get back to their command. It seemed that no other possibility occurred to them. Moro men brought the Navy fliers to Sitangkai, the westernmost island in the archipelago, from which the Americans took an old diesel-powered launch across open water to Dutch military headquarters in Borneo. They continued south to Surabaya in Java by boat, plane, and ship, following the retreat of the U.S. Pacific Fleet. On January 12, 1942, Christman walked into U.S. Naval headquarters and reported for duty. In the eyes of the U.S. military, this act brought him back to life.

GARBAGE

The picture of the albatross carcass bursting with plastic appeared in *National Geographic* in October 2005. Beside it ran the photo Susan Middleton took of the pieces she extracted from the bird's stomach, arranged on a white backdrop in an oval. "Like an egg," she said someone once told her.

A few readers wrote to point out that a white shard in the lower right of the photo said VP-101, the name of Christman's WWII Navy squadron. The magazine assigned a researcher to find out more. Louis Dorny, a naval historian, called the plastic piece "typical" of Navy ID tags attached to equipment, like a toolbox, or a bombsight.

Bombsight. The word fuses entities, monsterish. As if the bomb itself had sight and could seek its own target. Or it gave the bomber power to see like a bomb. Which would be what? A falling, blurred, a blast, maybe light and flame, and then what?

The bombsight in fact required the bomber to see the target with his own eyes. Then he could enter altitude, speed, and coordinates, and the machine would release the bomb at the right moment. It was supposed to give airplanes the power to destroy small things on the ground from very high in the air.

It never worked well, so pilots had to dive close before dropping their "eggs," which is what Christman called them in a letter home to his family: *Next week we are going to drop some 100 and 500 lb. eggs with T.N.T. guts.*

> how the egg of a bird is crystalline, made of layers lined with minuscule air canals so the chick inside can breathe; how the thickness of each egg's shell meets exactly the pressure each incubating bird will bring to bear

The plane that Christman flew to Jolo Island in the Philippines on December 27, 1941, carried three five-hundred-pound eggs. So did each of the five other airplanes.

Dorny considers the disastrous bombing raid that followed "the first, most obvious" incident that sent VP-101 equipment into the ocean. As he puts it: "Many men lost, and mass confusion...lots of loose ends."

Lots of loose ends. I never saw the photo of the albatross chick in *National Geographic.* I first encountered the piece of plastic in a 2006 *Los Angeles Times* article by Kenneth Weiss, who devoted two sentences to it. It stayed in my brain, this relic from WWII found inside a dead bird some sixty years later. This persistent little bit of death in life. It became a shard that stuck.

But I ignored it. I was writing a book at the time about a cemetery. We were trying to have a baby. And I was sick, I had to finally admit, sick from fear. Sick from fear and sick from grief at life in New York, a city with a funeral pyre at its core, in a nation thirsting for blood. I was sick from years of trying to ignore the terror—which comes from the root word for tremble—that had lodged itself in my body the morning I sat on the subway below the World Trade Center. I felt the train shudder and heard a noise and wrote "a thing that flies" in my notebook. A thing that flies. That beautiful September morning.

I stayed there and swayed in my life between the *no* and the *yes*. Mostly *no.* Mostly *no* was winning. I spent all my time in a cemetery.

Then Jen and I moved cross-country and agreed to give up the idea of a baby. As Claudia Rankine writes, rephrasing César Vallejo, "any kind of knowledge can be a prescription against despair." I started to think about this plastic bit. I wanted to know more. My observer grew curious again, roused herself to life. I contacted Kenneth Weiss, the reporter. He directed me to Curtis Ebbesmeyer, the oceanographer. Ebbesmeyer sent me to Dorny.

"Lots of loose ends," said Dorny.

Ebbesmeyer makes this guess: The piece of plastic got sucked into the Kuroshio current off the Philippines and spent sixty-some years circling the North Pacific before an albatross plucked it out of the water, perhaps with nutritious flying-fish eggs attached, to feed to Shed Bird, its chick.

It is the oldest piece of plastic from the ocean to which Ebbesmeyer has been able to assign a rough date. He and Dorny both presume it was made of Bakelite, but that is another loose bit: The piece of plastic has been lost.

This is what Susan Middleton, the photographer, reported to me in an email: "With regard to the bird's stomach contents, the last time I saw them was when I was carrying them around in a baggie and showing them in conjunction with presentations I did in Hawaiian schools in early 2006...On Moloka'i I realized I did not have the baggie. I contacted the school but after a search nothing turned up. It seems that perhaps the janitor disposed of the bag, assuming it was garbage."

"Garbage." The word once meant the waste parts of an animal.

A thing that flies. I wrote the phrase in a notebook just before coming above ground to find the World Trade Center in flames, paper and debris cascading. I wrote it before I had any idea that the tremor on the subway train came from the impact of an airplane running into the building above me. I found the phrase again months later, looking through old notebooks for poetry material. I had forgotten completely I wrote it just before. The moment that can now be called "just before."

The notebook is gone now. I can't find it. It's not in any boxes of the journals I have kept since high school.

Lots of loose ends. There is no way to know the true origin of the piece of plastic inside the albatross. The VP-101 squadron existed from 1940 to 1944. It lost planes off the Philippines, Indonesia, and Australia, as the Japanese pushed the squadron south along with the rest of the Pacific Fleet. The piece disappeared somewhere within a vast battlefield of water.

Maybe Christman rubbed this bit of stamped plastic in his fingers as he sat waiting for his crew to fuel the plane and load the bombs. Maybe it hung from the life raft that burned so badly they couldn't use it. Or maybe it never had anything to do with Christman, his crew, or his airplane.

The connection is, in that sense, false. I made it up.

But how else to exist? How else but to weave our frayed ends together, to weave and reweave the net? How else to live, which means "remain, continue"?

When we packed to move cross-country, I decided to get rid of them. All the journals for twenty years. I put them outside, four boxes, by the trash on 23rd Street in Brooklyn. That night it rained. I lay in bed and thought of them, the notebooks, outside getting wet. I got up and carried them back. They moved in the truck cross-country. But not that one, from fall 2001. That one is gone.

Maybe I threw it out in some kind of fugue state. I have always associated the word "fugue" with death, with funeral. Now I learn it means flight, the act of fleeing. A thing that flies. Airplane or albatross. A person.

The notebook has maybe disintegrated by now, or it sits entombed with rotted food, plastic take-out containers, clothing, and all the other objects discarded by the residents of New York City. And what about the albatross chick? Susan doesn't know what happened to its body after she took all the plastic out. It decayed, she imagines, in the hot Kure sun.

"Flies, ants, and ghost crabs consumed the flesh in short order," she wrote me, "eventually exposing the skeleton, and then that too, disintegrating...lots of calcium carbonate on Kure, from bones...coral...algae creating all that white sand."

Dissolved also: the airplanes that sank in seawater, the bodies of Christman's crew mates Pettit and Waterman. Or maybe their bones remain out there sinking slowly through sediment layers on the ocean floor.

What still exists, we know this—the plastic bits from inside the bird's body. Maybe they sit in the landfill on Moloka'i. Or maybe they're back in the water, washed down a storm drain or blown from a trash bin. The plastic will outlast the bones, the sand, this writing.

REFUSE

Decay means this: to turn into food for something else. Tiny living beings break the bonds that hold a substance together, transforming it into simpler molecules they can use for fuel. These creatures are called decomposers.

Paper presents a challenge for decomposers. It consists of cellulose, long chains of carbon, hydrogen, and oxygen that make the tree strong enough to stand upright. Only certain creatures (bacteria, fungi and single-celled swimmers that are neither) have enzymes that can break apart these complex chains and turn them into sugar. They live in the guts of termites and ruminants. They move through water and soil.

Molecules that form in long chains are called polymers, which means many parts. Cellulose, skin, hair, silk—all polymers. So is DNA. And lobster shells. And protein. The first plastic (celluloid, ultimately used to make film for movies) came from cellulose, the most common polymer on Earth.

The first synthetic polymer made in a lab using heat and pressure was Bakelite. Its developer, Leo Baekeland, studied with a protégé of Kekulé's at Ghent. Another Belgian chemist inspired him, Kekulé's friend and former student Adolf Baeyer. In 1872, Baeyer reported that he had mixed phenol with formaldehyde gas and produced a "hard, porous and insoluble grey substance." Baeyer and other chemists considered this an annoying byproduct because it hardened in test tubes and ruined them. Like Perkin, who invented mauve, Baekeland paid attention to what other chemists rejected.

Baekeland was already prosperous when he created Bakelite. He had developed the first practical photographic printing paper called Velox, an invention bought by Eastman Kodak. The paper could be exposed under artificial light in controlled conditions, instead of in sunlight, and it replaced the earlier albumen papers that relied on egg white to fix the chemicals, a technology that increased global demand for bird eggs and helped drive the slaughter of albatross.

Baekeland used his Kodak money to buy a house on the Hudson River in Yonkers, New York, and set up a laboratory in the barn. He looked for new problems to solve with chemistry.

He combined phenol, formaldehyde, and other chemicals, adding acids and bases, and varying the heat and pressure. He ended up with a material unlike any that existed: polyoxybenzylmethylenglycolanhydride, a matrix of linked benzene rings. It could be molded into anything; it set quickly, which was important for mass production; and it didn't conduct heat or electricity, so it was useful for the new electrical and automobile industries.

In 1907, Baekeland applied for a patent for this new material. He named it Bakelite and trademarked it with an infinity symbol, because it could take any shape imaginable. Soon Bakelite appeared everywhere: in buttons, bracelets, toys, telephones, radios, pipe stems, and guns.

Baekeland sold his corporation in 1939 to Union Carbide, which would become notorious for the world's worst industrial accident. In 1984, in Bhopal, India, a leak of methyl isocyanate, used to make pesticides and plastics, killed and injured hundreds of thousands. In 2001, as chemical companies around the world began to consolidate, Union Carbide came under the control of a new "parent"—Dow Chemical. This made Dow the second largest chemical company in the world behind DuPont.

Today, a handful of global corporations make nearly all industrial plastic using laboratory-fused polymers. Most decomposers don't recognize these lab creations as food; they lack any way to break them apart and absorb them into their bodies. So plastic becomes refuse: "rejected thing, waste material, trash." The word comes from an ancient root that meant “to pour”—the same root for the German word *giessen* that Heidegger uses. It means to pour back, flow back.

The rejected remains as fact. It flows through soil and water and air. It floats on breath and washes through cells. It fails to disappear.

ADVENTURE

eely things water weedy things

will be thoughts

—Alice Notley

Stan Ulam at first felt out of place in Los Alamos, a mathematician among the physicists. The long formulae scrawled across blackboards scared him. But he soon got a grasp of the laws that predict how objects interact. He wrote: "I found out that the main ability was to have a visual, and also an almost tactile, way to imagine the physical situations...Very soon I discovered that if one gets a feeling for no more than a dozen...radiation and nuclear constants, one can imagine the subatomic world almost tangibly." He called this approach "intuitive common sense," a "gut feeling."

To see as if

to touch. To see, an inside

sense, a sort

of felt thought. *Ad-*

venture from its

root—a thing on

the verge, about

to arrive—

Ulam's first assignment at Los Alamos: imagine how to spark a thermonuclear fire for the first time on Earth. Scientists had recently figured out how such reactions power stars. Only the fusing together of atomic nuclei could create enough energy to burn at such high temperatures—millions of degrees—and sustain such long-lived fires.

The force that fuels stars could create a "super bomb" here on Earth, a weapon of potentially unlimited explosive force. The problem: it would need an atomic bomb to get hot enough to ignite. Such a thing did not yet exist; the fact that an atomic nucleus could be split had been demonstrated only a few years earlier. The endeavor remained theoretical.

Edward Teller led the effort to think a super bomb into existence. A physicist from a wealthy Hungarian family, he ended up at a mesa top in the U.S. Southwest (like Ulam and so many others) in the wave of displacement driven by violence in Europe. Ulam describes him: "When I first met Teller, he appeared youthful, always intense, visibly ambitious, and harboring a smoldering passion for achievement in physics."

Anti-Semitism pushed Teller out of Hungary at age eighteen. He went to Germany, the center of the scientific world, where chemistry reigned supreme. Thanks to Perkin, Kekulé, and others, chemistry had risen from an obscure art, the legacy of alchemists, to a field that performed real-life miracles, transmuting the gunk left from burning coal into drugs and plastic and fertilizer, and making its inventors rich.

Teller studied the chemistry of polymers, but the strange problems of physics pulled at his curiosity. Physics would soon eclipse chemistry as the science of wealth and war, but in the twenties it remained an esoteric backwater. Like Perkin before him, Teller needed permission from his father to switch to an arcane field of study with an unclear future.

By the time Ulam arrived in New York in fall 1939, Teller was at Columbia, working with other European refugees to make the atomic bomb a reality. His colleagues were Enrico Fermi from

Italy, who had just won the Nobel Prize, and Leo Szilard, Teller's fellow Hungarian. Fear drove them. A lab in Nazi Germany had discovered fission; Hitler might be first to an atomic weapon. The foreign physicists wanted to warn the U.S. President about this new discovery regarding invisible particles, but they knew nothing about the workings of government.

They knew Albert Einstein though, and they thought he could get Roosevelt's attention. Szilard convinced Einstein to write the president a letter on August 2, 1939, alerting him to the possibilities of atomic fission. The histories often represent this document as a pivot point, a dark message laden with fate above a famous signature.

In fact, the words Einstein put his name to made almost nothing happen. It took months for the letter to reach Roosevelt, and more months for anyone in power to respond. Meanwhile, scientists across the U.S. (and in Europe, Japan, and Russia) kept smashing atoms to find what hid inside them.

In the midst of this, Teller found time to untangle problems in chemistry that intrigued him. One was the structure of the benzene molecule. In 1940, he and two others used a series of prisms and mirrors to direct beams of light through benzene gas. The experiment is called spectroscopy. It uses light to illuminate the invisible. Electrons absorb some of the energy, changing the light's intensity and giving clues to their positions within the molecule. The results provided evidence for Linus Pauling's strange theory that the electrons in benzene never occupy a single space but keep constantly moving.

Scope means "to look." Spectro comes from specter, "image, figure, a ghost." Both words stem from the same root, *spek*, "to observe." Spectroscopy: to look at the ghost that looking creates.

: :

Linus Pauling was born and raised in Oregon. Pauling's father died in 1910, when Pauling was nine and his sisters were eight and six. To support them, their mother ran a boarding house at the corner of what is now Southeast Hawthorne and César E. Chávez Blvd. in Portland, a few miles from me. I have driven by the house at this busy intersection countless times. One day, I stopped to look around.

It was a rather run-down beige three-story house with a broad front porch, flanked on one side by a tattoo parlor and on the other by a vacant business. Out front, a painted sign said, "Linus Pauling House, Center for Science, Peace, and Health." There was also a sign for Teras Recovery Couseling.

After watching me peer through the windows for a while, Terry Bristol let me in. Bristol is a retired philosophy of science professor, a native Oregonian like Pauling, who founded a nonprofit that aims to keep Pauling's legacy alive in Portland through a lecture series and other activities. To pay the bills, he also rents the Pauling house to nonprofits like the counseling center.

Bristol offered to give me a tour. The house, built between 1901 and 1905, showed layers of generations of occupants. Off the entryway was a curved staircase, a kitchen directly ahead, and to the right what would have been the parlor when wealthy people lived here, before Pauling's time. It had high, coved ceilings, brass fixtures, and dark wood trim lining the doorways and windows. Someone had laid cheap linoleum over the kitchen floor and installed a folding plastic partition between the entryway and parlor. Off the parlor was the formal dining room, now offices for the counseling center. Bristol pointed out the little alcove—where the sideboard would once have sat—that became Pauling's mother, Belle's, tiny sleeping quarters. She grew unwell after her husband died and was too sick to climb the stairs in the house, Bristol told me.

He took me downstairs to the basement, where Pauling set up his laboratory as a boy. The small corner where Pauling worked, separated from the rest of the room by a wooden partition, was full of boxes and supplies, storage space rented to the float tank business up the street, where people can lie in salty water in absolute darkness and silence. "This is our sacred space," Bristol told me, waving to the crowded corner. "This is one of those places on Earth that emanates a special energy." I asked him how he could tell. "Oh, I just say so," he said, with a little hint of humor. Bristol told me that to keep his two little sisters from bothering him, Pauling hung a sign outside his laboratory that said, "No girls allowed."

Pauling is the only person to date who has been awarded two Nobel Prizes not shared with anyone else. He received the prize for chemistry in 1954, for his discoveries about resonance and the structure of molecules, and he won the Nobel Peace Prize in 1962 for his activism against nuclear weapons. Because he marched and petitioned against nuclear war, officials in the U.S. Government suspected Pauling of being a communist and revoked his passport for a short time during the 1950s.

On February 20, 1958, at the height of the Cold War, Pauling engaged in a televised debate with Edward Teller. Pauling called for a ban on nuclear testing, citing the dangers of fallout around the world. Teller, perhaps the nation's most public and ardent supporter of nuclear weapons, argued that only a large nuclear arsenal could keep Soviet tyranny from spreading, as it had already spread to his native Hungary, cutting him off from his family. "The world has become small," bemoaned Teller, who had fled the Nazis only to see this new global threat rising. "I'm passionately opposed to killing, but I'm even more passionately fond of freedom." His words closed out the debate.

: :

All of this lay far in the future when, on a sparkling fall day in September 1941, Teller and the Italian physicist Enrico Fermi strolled Manhattan after lunch. They were headed back to the laboratory at Columbia, where they were trying to spark the world's first nuclear chain reaction on the way to building an atomic bomb. An idea occurred to Fermi: He wondered aloud whether an atomic bomb could ignite thermonuclear fusion and create a weapon that burned with the power of the sun. Fermi was one of the few people on the planet to have made this leap. The idea startled Teller. Over the next few days he conducted a flurry of calculations. He returned to Fermi with bad news: impossible. Such a bomb could not be built. But the idea of an ultimate weapon—perhaps of ultimate protection—held him in its grip. He could not look away from it.

ODYSSEY

After I hit a dead end with the parts company in Portland, I decided to try calling Honda's corporate headquarters. Amy answered the phone. I explained that I was looking for information about a car part. She transferred me to the parts department. Someone named Leonard answered. I told him that I was writing a book about plastic, that I had this car part, and I was trying to learn the origins of it, where it came from. A long silence followed my explanation. "Are you looking for a part?" asked Leonard.

I gave him the part number. "That's an invalid dealership code," he said. He suggested I give the part number to a dealership, and they could look it up. "We can only go from car to part, not part to car," he told me. He seemed hurried, impatient. He told me again to call a dealership and hung up on me.

I called Ron Tonkin Honda here in Portland. Ron Tonkin is ubiquitous in car ads in the city. This time I skipped the story and just said I was looking for help with a part. I was transferred to Rocky in the parts department. I read off the part number. Rocky looked it up and told me I had the part number wrong. I read it wrong off the plastic and have had it wrong ever since. "It's SXO, not SXC" he said. It's a fender liner from a 1995–1998 Honda Odyssey, a first generation, Honda's first minivan.

Rocky told me that a Honda plant in Lincoln, Alabama is now the sole manufacturer of the Odyssey, but it did not open until 2001, so this part had to have been made in Japan. I explained that I was writing a book about plastic and I wanted to find out where this particular car part came from. Rocky said nothing at all. Once again, the silence stretched out. This time, I broke it, because I was afraid Rocky would hang up. "Do you have any idea where this car part might have been made?" Rocky said Honda in Japan still manufactures this part for older minivans, but he had no idea where.

It turns out that the legend of the Odyssey has been told and retold throughout the auto industry.

O Muse, tell me

of that ingenious hero

who traveled far from Tokyo

to build a car to satisfy

the creative lifestyle of the American consumer.

In 1990, Kunimichi Odagaki, chief engineer at Honda, received a phone call from the company's research and development center. They wanted him to create a minivan for the American market. Honda had never built such a thing before.

The American journalist Micheline Maynard described Odagaki in her 2004 book, *The End of Detroit*. "Modest and soft-spoken, with curly dark hair, Odagaki yields [sic] enormous power within Honda...He demands such respect he may ascend to one of the company's top jobs, perhaps even as chief executive someday."

When Honda issued its challenge to Odagaki, Chrysler, which invented minivans, dominated the market with the Town and Country, Dodge Caravan, and Plymouth Voyager. But Odagaki wanted to create something totally different. His concept: "personal jet"—three rows of seats with a center aisle. He gave it the code name PJ.

Odagaki assembled a team and flew to the U.S. They traveled across the country interviewing minivan owners and observing how they used their vehicles, following a Japanese practice that Maynard translates as "go to the spot." The team traveled 25,000 miles over six months. They drove American minivans so they could directly experience their problems and advantages.

They parked outside of elementary schools and videotaped families dropping off children (which got them in trouble with the cops). They sat in Home Depot parking lots to watch Americans engaging in the hobby of home improvement, which was strange at the time to these Japanese visitors. They visited homes and talked to minivan families.

Soon after Odagaki's trip to the U.S., Honda canceled the project. Japan's economic bubble burst and the stock market crashed—it wasn't a time for risky new ventures.

Odagaki decided to continue the project in secret. "I insisted... perhaps with a certain amount of self-praise," he remembered, "we could change the world by introducing this car."

To convince Honda's corporate managers, Odagaki had his team build a life-size Styrofoam model, which took up most of the conference room where they gave their presentation. As the executives climbed in and out of the foam car, the designers emphasized the van could carry four golf bags without having to stack them, and it would look good at the clubhouse entrance. That won over Nobuhiko Kawamoto, Honda's president.

Honda finally launched its world-changing minivan on October 20, 1994. Odagaki cried at the launch ceremony, saying that he believed adversity drove them to create a better product. "The team members had a clear idea as to what we should create," he said. "But we also knew how to fight."

By 1997, the Odyssey had become Honda's fastest selling new vehicle.

In 2014, twenty years after the original, Honda unveiled an updated Odyssey. It cost $45,000 and included a sixteen-inch video screen for the back seats, a built-in vacuum cleaner, and a cooler box. "You will lack for nothing," wrote the *New York Daily News* in its review.

You

will

lack

for nothing.

II.

REFUSE

REFUSAL

I attended my first writing residency in 2014, in the desert of eastern Oregon. In preparation for two weeks devoted only to writing, I sketched a map of this book, an attempt to get a sense of its themes and connections. I tried to make the map look something like a complex molecule, but I have no skill for drawing.

I packed clothes, food, the car part, and the map into the back of my car and drove six hours across Oregon, from the Willamette Valley and its green patchwork of farms, over the Cascade Mountains draped in Douglas fir and hemlock, to the eastern desert, dominated by ponderosa pine and juniper, like the Pajarito Plateau of my youth. During the five-hour drive, I sent about 230 pounds of carbon dioxide into the atmosphere, destined to stay there for centuries.

Summer Lake. We'd camped in this area before. It's one of my favorite places in Oregon. A painting of it hangs on my wall. The lake is large and shallow, a salt lake, the remnants of a vast Pleistocene water body that is mostly dry much of the year. It used to be a lush, fertile basin: The oldest known human remains in the Western Hemisphere come from nearby, dating back 14,000 years.

Now, in this age of settler colonialism, hardly anyone lives here. It's called the Oregon Outback. The parts of the lake that remain marshy and wet provide a haven for an astonishing array of birds, including sandhill cranes. I could hear them bugling in the mornings and watch them walk in their slow, jutting way through the grass. Wind blowing across the dry lake lifts the salts into strange, hanging veils—an ocean of light and shadow.

My cabin overlooked Summer Lake. It had a sliding glass door that opened onto a large deck. The glass was covered with reflective material to keep out the full glare of the sun. In the mornings, a quail family would come right up to the glass—the male with a fillip of black feathers on his head—bobbing and tilting at their reflections. The first day I was there I set the car part down on the carpet, spread out the map, and sat staring at it, wondering where to start, where to go next. I stared and stared—down at the map, out at the desert, down at the map. It struck me: The book was full of blanks—people I had erased. I'd been reading technical and historical texts: histories of plastic, of the atomic and thermonuclear bombs, World War II airplanes, European canonical writings. White men dominated these accounts. The histories entangled other lives, of course, but they disappeared in the texts, erased. I had repeated that erasure.

That night I had a dream about my niece, a dream so visceral I still can feel it five years later. She was just blooming into puberty, radiant, bursting with life. I brought her out to the pale grey mud bog in the center of Summer Lake, the place, I knew in my dream, where everything dies. A giant silver fish stuck out of the mud beside the path, only its head and gills showing. Arielle reached to touch it.

"No!" I told her, more harshly than I'd meant to, "too dangerous." I turned to point to the massive manta rays dying on their backs, white undersides flopping and writhing, dark gill and mouth slits. I turned back. Arielle had vanished, no sign of her across the dying flat. I screamed. A man in waders fishing nearby seemed unconcerned. He swiveled his head toward me then swiveled it back. I screamed again and the ground before me started to bubble and shift, her pale hand shot up. I grabbed it and hauled her out, wiping, so she could breathe, the mud that caked her nose and mouth.

It was a birth dream of course, but it came with a clear understanding: Arielle had given birth to herself. She'd done something risky—stepped off the path to touch that fish—but she saved herself, got herself to the surface and thrust up her hand for me. I wrote a poem about the dream the next morning. I ended it this way:

We start out dead

as women, objects already

in a dying world, seduced

by its glitter to add ourselves

to the museum of desire forever.

To live, Arielle,

you will have to disobey everyone

all the time, not once

but your whole life,

over and over. Disobey

everyone else, give

birth to yourself

I sat on the deck of that beautiful place feeling surges of anger. It came like waves, building with the salt winds off the lake—anger at the silences I accepted and repeated, anger at the silence I sealed inside myself. At the caution and vigilance I keep, the accommodations I make.

A jeweled hummingbird flashed toward me; it buzzed up under my chair. Startled, I leapt out of my seat. Hummingbirds give me the creeps. Aggressive, territorial, with their metallic, scissoring sounds—the bird's buzz vibrated to my anger, a tear in the light.

I wanted the defiance, the birth I'd given to Arielle, for myself. I felt rage at people in power for performing the theatre of doing something about the burning planet while taking no action that might threaten the economic order. I hated writing this book. I hated picking up plastic, seeing it everywhere I went.

The literary companions I carried with me on this trip included Alice Notley and Audre Lorde. I brought them in around me: Notley flying with owl sight and predatory wit; Lorde drawing female power from Earth, from the dark of her felt thought. I wanted to burn everything up. Underneath my anger, though, ran a deeper seam: grief. I stepped out from inside myself, into the place these poets opened for me, and I started writing. Poems poured out of me—songs of lament and sorrow.

They became *After We All Died*, a different book, but still haunted by Heidegger, and a cemetery, and nuclear fathers—my own refusals flowing back to me.

THE COMPUTERS

On December 1, 1949, Heidegger delivered his lecture "The Thing." Three months before, on August 29, the Soviet Union exploded a plutonium bomb in Kazakhstan. The rest of the world only learned about the Russian weapon when, on September 3, a U.S. Air Force reconnaissance plane above Russia's Kamchatka Peninsula detected radiation three hundred percent higher than usual. Scientists tracked the radioactive cloud as it moved across the Pacific Ocean to North America. It hung over Washington, D.C. The Naval Research Lab collected radiation in rainwater off its roof.

News of the Russian bomb surprised everyone. Many at first did not believe it—not Ulam, the mathematician, not President Truman, not the head of the military, Louis Johnson. The men felt their cloak of atomic protection dissolve with the radioactive rain falling on Washington. They would need a more powerful weapon.

As Heidegger conjured the hydrogen bomb in his lecture and its potential to "snuff out all life on Earth," Ulam, Teller, and others at Los Alamos struggled to make such a weapon real.

It took a lot of math. Designing a bomb of unlimited power does not proceed by trial and error. Instead, the scientists had to simulate with numbers the process and conditions under which a fission reaction could generate enough heat and pressure to fuse a nucleus together, and another, and another, sparking a thermonuclear fire.

The mathematicians needed to capture the whole sequence of events, tiny particles moving through tiny increments of time, one ten-millionths of a second, like freeze frames in numbers. Ulam called it the largest mathematical effort ever undertaken.

This did not deter him. As Heidegger completed his series of lectures in Germany, Ulam and his colleague Cornelius Everett began to compute the numbers. "We started to work each day for four to six hours with slide rule, pencil, and paper," Ulam wrote. "It was long and arduous work."

To help, they enlisted "a bevy of young women," in the words of Ulam's wife Françoise, herself among them. Wives of scientists often served as what were called "computers." Some had college degrees in non-science fields, giving them few job prospects in Los Alamos. Bevy is fifteenth century, "a collective noun of quails and ladies." Of unknown origin, perhaps from an Old French word for drinking bout, as in a company of birds gathered at a pool or puddle for drinking or bathing.

"We bore the glamorous name of 'data analysts,'" Françoise wrote. The women received little information about the purpose of their work. They spent hours bent over bulky metal calculators, punching in strings of numbers. Machines would soon replace them.

A few years earlier, one of the first digital computers blinked to life. The Electrical Numerical Integrator and Computer, ENIAC, weighed 30 tons and contained 18,000 vacuum tubes.

The problem with the ENIAC: it had no central processing unit. It had to be physically reprogrammed for each calculation by adjusting its three thousand switches and circuit wires, an intricate process that required weeks of effort. Women also largely performed this labor. As the women at Los Alamos crunched numbers on calculators, other women programmed the ENIAC to perform the same work.

Common wisdom held that women were better at this type of tedious, repetitive labor, because of their "constant alertness, nimble fingers and tireless wrists."

Dip and drink dip

and drink birds bent-

necked on constant

alert one

black eye turned up.

In fact, programming demanded innovation—it had never been done before—and an intimate knowledge of the machine and of the math involved. The women learned the ENIAC by studying blueprints, then by climbing and crawling around its plug boards and function tables. They knew the machine better than the engineers who built it. "We could diagnose problems almost down to the individual vacuum tube," said Jean Bartik, one of the original programmers.

The people at Los Alamos felt an urgency for answers. The world once again seemed a darkening place. The Soviet Union consolidated its control of Eastern Europe with a coup in Czechoslovakia in February 1948. In the summer, Russia blocked Western access to Berlin. In the spring of 1949, the communists took over Hungary in a sham election, a development that cut Edward Teller off from his mother, father, and sister, who had survived the war and remained in Budapest. That August, the Soviet Union exploded its atomic weapon. In October, the communists declared victory in China.

Responding to these threats, President Truman announced a crash program to build a thermonuclear weapon, which the scientists called "the Super." Teller returned to Los Alamos from the University of Chicago to work on the weapon that had obsessed his thinking since Fermi first suggested its possibility in 1941.

By December of 1949, Teller's design appeared dead. The results from the human computers suggested that it would fizzle, it would not ignite a thermonuclear fire. The ENIAC confirmed their calculations.

Teller directed his frustration and disappointment at his colleagues, with outbursts of temper and demands for more resources. He feared Ulam might be trying to undermine him. "I sensed the tension in the corridors," wrote Françoise.

The computers provided a diversion for the men. Ulam wrote:

> I particularly remember one of the programmers who was really beautiful and well endowed. She would come to my office with the results of the daily computation... She would unfold them in front of her low-cut Spanish blouse and ask, "How do they look?" and I would exclaim, "They look marvelous!" to the entertainment of Fermi and the others.

In addition to the wives of scientists, other human computers came to Los Alamos as recruits to the war effort. Many remained anonymous, a "room with girls in it," as the physicist Richard Feynman put it.

The six original ENIAC programmers also disappeared from history. The women all married after the war, which transformed their identities, making them harder to trace. Only a few, like Betty Snyder (later Holberton), stayed in the industry. She developed much of the software for the first UNIVAC computer and helped create the early programming language called Cobol.

In 1946, the War Department publicly announced the ENIAC, a new "intelligent machine." It made no mention of the female programmers, and neither did a *New York Times* story on the new computer. The *Times* ran a photo that became one of the most widely reprinted of the ENIAC. It featured a man in uniform plugging wires, with two women barely visible in the background. When the Army used the photo for postwar recruiting, it cropped the two women out.

For the armed forces, the decision to crop women out of postwar recruiting materials made sense. They considered women temporary replacements. A War Department brochure from 1943 titled, "You're Going to Employ Women," makes this point. The brochure provides practical advice for hiring, training, and managing these strange creatures. It includes tips like "interpret machinery operations in terms of household and kitchen appliances," and "use a trained personnel woman" who can "give sympathetic attention to home problems."

The brochure ends with this exhortation: "Remember... A WOMAN WORKER is not a man. She is a substitute—like plastics instead of metal—"

But plastic did not go away once the war was over. As the brochure admitted, women, like plastic, had "special characteristics that lend themselves to new and sometimes much superior uses." In women, those traits included being "pliant, accurate, dexterous, and good at repetitive tasks."

The benefits of plastic were less clear, but potentially infinite, as the original Bakelite trademark symbol implied. James W. Sullivan, an executive at Union Products—best known for the pink lawn flamingo—recalled that immediately after the war, "Virtually nothing was made of plastic and *anything* could be."

With wartime funding and demand from the U.S. Government, plastic producers swiftly developed new materials, new processing techniques, and new abilities to meet exacting specifications for military equipment. Plastic production tripled between 1940 and 1945.

The demand came not only because of a shortage of metal and a need for new materials. Manufacturers encouraged it. In 1941, the Society for the Plastics Industry created a Plastics Defense Committee. Its job: to promote plastics for wartime use, as a "young, virile industry unhampered by ancient ideas and stodgy procedure."

The result of this government-industry alliance: plastic found its way into thousands of military uses. These included commonplace items like raincoats, combs, razors, and the equipment tag that ended up in Shed Bird's belly. It also encompassed the highly specialized: the M-52 mortar shell fuse, barrage balloon release valve, and rocket launcher tube. The atomic bomb apparently used Bakelite, but exactly how remains classified.

All this industrial capacity needed new outlets after the war. Advancements in oil refining meant that refineries could operate in a "continuous flow," producing gas, oil, and raw ingredients for plastic twenty-four hours a day, every day, without stopping. Shortly after the war, improvements to injection molding made it easier for manufacturers to shoot liquid plastic into premade molds, spitting out thousands of identical items "faster than we can tell you about it," crowed *Monsanto Magazine* in 1947.

The stuff pouring out of factories threatened to become junk before it could be sold. This is the reason the DuPont executive in 1945 exhorted marketers to "see to it that Americans are never satisfied." This gave rise to a sophisticated advertising industry aimed at creating "consumers"—people who looked to purchase things to meet needs or solve problems often invented by the industries themselves. The goal was to construct a shopper "plagued by inadequacies and threatened by his or her own failings." Advertising helped people construct imaginary selves, the ideal a person might reach by acquiring the right products.

But even this was not enough. One of the biggest problems with plastic—originally touted as one of its greatest benefits—it lasts forever. People did not have to buy a plastic garbage pail or Tupperware container over and over. So the industry switched tactics. In the 1950s it began to focus on single-use items like Styrofoam coffee cups, individual jelly packs, and polyethylene bags for dry cleaning.

Still, people had to be taught to throw plastics away. Those who lived through the Depression and war were accustomed to conserving. They saved their plastic cups for reusing. At the national Society for the Plastics Industry conference in 1957, an editor with *Modern Plastics* urged manufacturers to develop products destined to wind up "in the garbage wagon."

The best way to do this? Focus on a product with an inherently short life: packaging. Takeout trays; egg cartons; shampoo, bleach, and detergent bottles; six pack rings—all served only as long as the consumable items they contained. The creation of the blister pack in the 1960s—a clear plastic bubble to protect and display the contents—made it possible to package almost anything in plastic.

By 1969, nearly a quarter of plastic went into packaging to be used once and then discarded. Today, that number is forty percent, by far the largest single use of plastic globally, and the largest component of plastic waste.

Its lasting power means that essentially all the plastic created still exists, except for the small amounts burned up in garbage incinerators or bombs. Also, bacteria in the ocean and some species of fungus may be eating it; certain beings seem to have evolved this ability. Still, most of the plastic ever made remains with us, circulating through water, living bodies, and the atmosphere—and the waste keeps coming. The world tosses out about 300 million metric tons of plastic every year, nearly the weight of the entire human population.

To consume means "to destroy by separating into parts which cannot be reunited, as by burning or eating." In that sense, we humans do not consume plastic nor much else produced by the industrial economy.

Instead, most consumer plastic—the packaging—serves as the vehicle of delivery for something else, something made to satisfy desire, soothe fear, assuage grief. The plastic is the shell of that longing: the shells, billions and billions of them collecting all over Earth, the way the discarded shells of purple-dye-yielding sea snails piled up on the shores of ancient Tyre. It is the trace of the lust for something else.

LAMENT

In 2014, I saw my first albatross. The poet, professor, and publisher Susan Schultz invited me to the University of Hawai'i at Mānoa to read ecopoetry. I flew, once again, 2,592 miles across the Pacific, blasting out planet-warming gases as I went. This visit marked my place in a long lineage: white settlers imbued with authority to speak as protectors and defenders of the ecologies they helped wreck.

The word "complicit" comes from a root that means to fold, or weave together. I stood before students and colleagues in Hawai'i, threaded to them by the violence I carry, in my ancestry, in my body, in my every step and breath. Other threads also tied us together—the care and concern we shared.

"Concern" comes from a root that means to sift or sieve, plus "con," the root for together. To sift or sieve together. It suggests a mixing in which the constituent parts retain their integrity. A weaving that displays each original color. This mix of students, poets, teachers—all of us in this place with our particular heritages, our histories, our unequal suffering, our stakes. It was our care—whose earliest meaning was mourning—that held us there.

We spoke of the plastic washing up on the islands, and we collected it. The campus seemed pristine, every inch mowed and manicured, but we found the plastic easily, tucked under bushes and blown into corners. The discarded food wrappers swarmed with what I thought at first were beetles, but quickly realized were the cockroaches that thrive everywhere here, another plague brought by human settlers.

Albatross nest mostly in the remote Northwestern Hawaiian Islands, but I learned there is a small nesting colony at Ka'ena Point, on the northwest tip of O'ahu. The birds arrive in late fall and nest through early summer. I came in mid-November.

O'ahu's birders had not yet seen any albatross, but we took the chance anyway. On the last day of my visit, Susan and I and the two other visiting poets, Stephen Collis and Brian Teare, hiked out to the point.

We walked three miles down a dry, flat track along the volcanic shoreline, the sun beating down on us. We entered through a double gate in the predator-proof fence and approached the grassy rise where the birds nest. We could see nothing. Brian and I scrambled off the trail up the hill, and there at the top, not five feet from us, sat an albatross in its nest, its dark back to us, white head tucked. We could see a few others nesting in little depressions across the hilltop, white heads and dark wing feathers visible in the grass.

We lay on our bellies, looking. I wanted to be amazed, seeing this bird for the first time, this creature that spends almost all its life in flight. But I couldn't focus. It felt like a violation to be so close, and we worried about disturbing the birds, so we slithered back down the hill after just a few moments.

In reality, we could have come up and thunked the bird on the head. A nesting albatross will not move off its egg.

The year after I visited Hawai'i, wildlife managers found mice on Midway Island eating nesting albatross birds alive on their nests. Mice and rats, introduced to the island by the U.S. military, have long tormented albatross. The scientist Carl Safina writes about birds with their internal organs showing, half chewed apart, near death, still sitting.

The rodents are only the latest in a long history of human-caused threats to albatross. Safina gives a sickening litany. In the seventeenth and eighteenth centuries, European sailors on imperial adventures fished the strange birds out of the sea and teased and tortured them onboard, or shot them for sport, the practice that inspired Coleridge's poem.

In the nineteenth century, hunters gathered rail cars full of eggs used by the new photographic industry. They also exterminated adults for feathers, which sold as "swan's down." Safina recounts that a visitor to Midway in 1902 found great heaps, waist high, of albatross that had been killed with clubs, their wing and breast feathers stripped. In 1909, President Theodore Roosevelt declared the entire Northwestern Hawaiian Islands—an outpost of the American empire—a federally protected bird sanctuary. Poachers came to Laysan Island two months later and killed 128,000 albatross.

Midway Island is halfway between Japan and the U.S., and the U.S. Navy established a base there in the lead-up to World War II. Jack Coley, a crew member on a PBY airplane like the one flown by Christman, recorded the Navy men's interactions with albatross, which they called "gooney birds." He wrote that they played games with them during long, hot days on the island, chasing the birds crosswind as they tried to fly so that they couldn't get enough wind support to take off. The air crews lived at close quarters with the birds and got to know them intimately. Describing the albatross mating dance, Coley writes, "Their motions and movements...caused many [a] PBY pilot and crewman to...mimic them. And later, in the states or at a forward base...these men became the center of attention demonstrating the 'Goony Bird Dance.'"

But the military establishment found the birds a nuisance and tried to eliminate them. "They bulldozed incubating birds, and at one point even used flamethrowers," writes Safina.

In December 2015, a group of teenage boys on O'ahu entered this lineage. They went to Ka'ena Point two days after Christmas with a baseball bat, a machete, and a pellet gun. They climbed the same grassy hill we had climbed and began clubbing, shooting, and hacking at birds. They killed fifteen albatross, destroyed nests, and smashed eggs.

The killings shocked many on the island, and the news spread far beyond Hawai'i. The boys attended the prestigious Punahou prep school, Barack Obama's alma mater. Wildlife biologist Lindsay Young waited nine days and then went to the point to gather the dismembered body parts by following the scent of rot. She directs Pacific Rim Conservation and has studied the birds at Ka'ena for years.

Only one of the students, Christian Gutierrez, was eighteen at the time of the attack. Charged originally with felony theft and animal cruelty, he pled no contest to reduced misdemeanor charges. On July 6, 2017, before a packed courtroom, Young wept as she spoke from the witness stand about gathering dismembered feet and beheaded bodies. She said the killings left her "life's work and spirit shattered." A judge sentenced Gutierrez to forty-five days in jail, and he walked out in handcuffs. His lawyer expressed hope that Gutierrez could serve his sentence and get back to his classes as a film student at New York University.

The deliberate destruction of albatross today seems rare, but human damage to the birds is vast, greater than it's ever been. Fishing kills an estimated hundred thousand every year. Industrial boats tow fishing lines that stretch for as long as eighty miles, baited with thousands of hooks targeting tuna and swordfish. Albatross dive for prey on the lines, get snared, and are dragged underwater to drown.

Another image from Safina's book haunts me: an albatross with a hook in its bill being pulled straight down through the water, its enormous wings bent in half. He estimates that albatross worldwide have declined about forty percent since the 1960s. But he does not give up hope for the long-lived, slow-breeding birds. Safina points out that albatross are increasing where their nesting sites are protected. Some countries and fisheries have adopted simple techniques to keep birds off the long lines, like using bright-colored streamers to scare them away and setting lines at night when the birds don't hunt. It remains a challenge, though, to enforce these practices on fishing vessels that cross the remote ocean.

In his book, published in 2003, Safina does not mention plastic as a cause of death for the albatross. Recent research shows nine of ten dead albatross in the Northwestern Hawaiian Islands contain plastic in their stomachs, and the same is true for seabirds worldwide. It also is true of humans. A 2020 study found invisible plastic particles suspended in the air and raining down everywhere the researchers looked, so much plastic they kept rechecking their results. They concluded plastic fallout exists in "every nook and cranny" of the planet. Like birds, people eat and breathe it, on average the weight of a credit card in plastic goes into people's bodies each week. Scientists also have found plastic particles in placenta that nourishes human fetuses. No one knows what this means for us—we all, the concerned, threaded together.

DESIRE

I started thinking of the car part as a creature, something like a bat, black with its folded wings full of corrugations. It ended up in corners, a dark, drooping thing, lurking. I stopped seeing it, or thinking about it for stretches of time, but I kept collecting plastic, still not sure what I was doing. I tried different approaches: arranging plastic as a collection from each walk and taking a photo, making a list of each item I found. I felt drawn to the language in plastic garbage: a yellow balloon scrap with a white bear that said, "Fred Bear." A little plastic tab labeled "out;" half a coffee cup lid that said "here;" a plastic bag labeled "roundhead screwhole plugs;" an eyedropper container branded "Refresh Plus;" a syringe that read "Easy Touch." Language of information, direction, marketing, a word to soothe the needle's prick.

It all felt a bit empty—as easy to gather and forget as plastic makers would like. I wanted to think about other words, words that lay underneath the language of marketing and manufacture, words that might explain, or even hint at, the drives behind such relentless waste. I started labeling pictures of plastic with my own words: "fast," or "fear." "Desire." I tried "greed," but that didn't seem elemental enough. It turned into "grief." I wrote a single word in marker on a piece of paper and stuck it into the photo with the plastic bit, or sometimes I wrote on the plastic itself. A blue pacifier: desire. Apples, bananas, and oranges at the airport individually wrapped in plastic: fear. Styrofoam packing material sticking out of a garbage can: desire. A condom: I labeled it grief.

Jen told me years later that she would come across these scraps of paper around the house and yard. She said it was a weird, dissonant experience: Where one might expect love notes, or grocery lists, or reminders, she found instead these disembodied nouns. *Grief. Fear. Desire.*

More and more as I labeled, I felt like I could make these words even more elemental. I could reduce them to a single one: fear. Fear of death. Fear of being alone. Fear of lack. Fear. Fear. Fear.

But this labeling also left me unsatisfied. It felt...I don't know. Too performative. Obvious. I had collapsed the specificity of waste—waste that might reveal something about those who possessed and disposed—into a uniform landscape of signifiers. It was pedantic. Monotonous. So was the waste itself. I framed things I found, with the camera and with words, in ways that caught my interest. In reality, most plastic crap is boring and relentless—cigarette butts, bottle caps, water bottles, food wrappers—the castoff bits of the numbingly endless stream of commodities.

In my wanderings, I came across at least three plastic car parts identical to the one shoved in a corner at home. I passed them by without stopping. They were too big to collect.

JOY

In June 2018, my mom died of lung cancer.

For most of my life, I disregarded and disrespected my mother. I learned this from the culture that surrounded me, from the town, with its emphasis on achievement in science, and from the larger society. She was a housewife, performing the menial labor of care, with no public status, no measure of worth. Frivolous talk of food and shopping. I could almost never bring myself to listen to her, or to respond in anything other than monotones when she talked to me. This started in middle school probably, when the girls in my grade began to bully me, and I realized she could do nothing to protect me.

But this represents the self-centered and partial understanding of a child. My mother protected me in powerful ways I saw and didn't see. She created me, her self woven all through my being. She taught me, in words and actions throughout my life, to love and value myself. She also—her death made clear—bound our family together and carried forward our traditions.

My mother always said she didn't want to linger in old age and dementia like her mother, her aunts, her grandmother. She smoked for most of her life and loved to eat and drink. A typical conversation with my mother involved a litany of what she had eaten or a recitation of a new recipe. My sister and I now grievingly and half-jokingly recreate these exchanges.

My mother had already survived breast cancer, with rounds of chemo and radiation. When she learned the lung cancer had spread to her brain, she wanted her death to come quickly. The day my parents called to share the diagnosis, she said, quoting Winnie the Pooh, "I am a person of very little brain, so this will not take long."

She did not want my sister, Paula, and I with her, or anyone except my father. "My husband takes care of me," she declared to the nurses and us. We persisted, and the nurses lent gentle pressure, and finally she relented. I paused my pursuit of plastic, paused everything, and went home to New Mexico. My mother refused to let us stay in the house. Paula and I slept at the home of a friend of theirs from church in White Rock, down "the Hill" from Los Alamos, really a mesa cut with sheer canyons that drop down to the Rio Grande Valley.

It was another season of devastating wildfires. The smell of smoke awoke me one night. I could hear my sister coughing in the other room. The next morning, we saw the white column rising from the Jemez Mountains above town. My parents had evacuated twice for fires over the years, and the fire filled my mother with anxiety about having to flee with her oxygen tanks. I noted the new weather forecast: "smoke." The sun hung, a red ball; the moon was bloody also.

But the fires stayed at bay, and each day we made the drive up the cliffs to our mother. All three of us, my father, my sister, and I, orbited around her illness, its powerful gravity. I rubbed her feet, legs, and arms, which were swollen and painful with fluids her body could no longer clear. I watched her sleep in her chair, thin plastic tubes feeding life-saving oxygen through her nose into her lungs.

At her insistence, we took her, oxygen tanks in tow, to get her hair done. The woman who had been styling my mother's hair for thirty years handed me an appointment reminder for five weeks in the future. We exchanged a silent look.

My mother planned her funeral, sitting with her pastor and my father. They commissioned a local potter to create urns, the same design for both of them. It was their fifty-fifth year of marriage. They planned to be cremated and set in the same niche in a wall beside their church, with a view over the valley to the Sangre de Cristos.

The pastor asked for all of our full names. The females on my mother's side of the family all have the middle name Mary, except for me. My mother broke family tradition to name me Kristen, after her best friend. When she did so, the story goes, her own mother wouldn't talk to her for a week.

My mother told the pastor this revised story: "When Allison was born, she said, 'I will not be named Mary!' So I named her Kristen." This confusion in my mother's mind startled me, but I also loved her story. Just as in my dream of Arielle, she gave me my birth and infused her disobedience into me.

: :

Three weeks from her diagnosis, my mom could no longer speak above a whisper. She couldn't really eat, though my father still insisted we all sit down together, and she could only pretend to sip from the wine glass we put beside her.

June 24, a Sunday—hard. Even the oxygen flowing constantly into her nose, even the painkillers, didn't ease the coughing and wheezing. She would cough and spit and then look at me, rolling her eyes, annoyed about it. We had given her a pan and a spoon to bang when she wanted something, a sort of joke, since one of us was always hovering.

I have a video of her from that afternoon, looking at me with arch humor, banging on the pan at my urging, but not wildly, delicately, controlled, with precision. She looks great, like herself, her hair well coifed.

She asked for lamb for dinner and we took her shopping; this was something my dad insisted on, getting her outside every day. It seemed almost a cruelty, it required so much for her to walk even a little, but she wanted to go.

We wheeled her into the town's single grocery store, her domain, and she bossed us both about what to buy. I had to lean down so she could whisper in my ear, translate what she said for my father, and then back again. It was fraying my nerves. "I have got to get out of here," I kept thinking. But then, her oxygen tank ran out, and she was gasping. "Take her out to the car and hook her up to a new one, I'm checking out," my dad said, and walked off with the cart. I couldn't believe this was happening. I ran her out to the parking lot. I'd never hooked up her oxygen tank because my father insisted on doing it all himself. But we managed. She didn't die there.

That evening, she couldn't taste the lamb, and couldn't swallow it either. She coughed and choked, disgust and disappointment on her face. My sister had gone home briefly to care for her own family, so it was only we three. The moments were agony. I suggested we pray, something I hadn't done for, oh, who knows. Decades. We held hands and my father led us. I knelt on the floor, between their recliners. My father prayed for her suffering to ease and made clear how glad and grateful we were that she was with us.

I knew she wanted to die. She wanted this to be over. When the prayer was finished, she turned to me and told me to express that to my father.

On my way out for the night, I leaned down and looked directly into her eyes. "Mom," I told her, "I'm going to pray for you to go, because I know that's what you want, then you can be free." Her blue eyes flashed at me. "*Yesss*," she whispered with all the power she could muster. As I headed out the door I called out, "Goodnight, Mom." She raised an arm to me.

My father called at around eight the next morning. When I arrived, she hadn't yet died, but I think she might have already gone. Anyway, she didn't respond to us. We sat with her, holding her hand, talking to her, praying, until we knew her breath had stopped.

: :

People when they die go to dirt. That's what I always thought. Or maybe some part of them continues as energy or something, that's it. I considered it naïve to believe that people remain in some form as individuals. I thought it might be a source of comfort, this belief, a psychological delusion that helps one survive traumatic loss.

She remained, though, my mother. My sister, who has always been attuned to such things, felt her first—a sense of total peace, Paula reported, as she was boarding the plane back to New Mexico. I dismissed this. I never really believed these experiences my sister shared. That night, though, sleeping alone, upstairs in the borrowed house, I felt it also. I felt her approach me. But I shut her out. Not by choice, some instinct. I don't know why. I didn't want her.

A few days later my father was driving us down the freeway to Albuquerque. He and my sister sat in front. I was in the back, the sun pouring in on my face. We were traveling along the Rio Grande, marked by its ribbon of green. He began a litany of all the ways he failed my mother: how he never bought a new washing machine, and she developed techniques that none of us now know for keeping it balanced in the spin cycle; how he let ice build up by the garage one winter and she slipped and broke her shoulder.

I felt her then with force—not speech but it was clear as speaking. She was shaking this off, his words, his regret, like a horse flicks off a fly, "Pah! It's nothing." A shake and then a leap.

"I feel just joy from her." My sister said this gently to my father. But I felt, speeding down the freeway, something else—I felt impatience from her, fire—a leap and then a crackle. *Catch!* it said. She—or it—wanted us to feel this too—to leave our grief for joy, catch up with her.

LOSS

The effort to remember the women, to re-enflesh them in the history, it comforted me. It also deepened my grief. And my anger. Everywhere their trace felt threatened, on the verge of total erasure.

There was Ida Noddack, a German chemist who suggested in a paper in 1934 that an atomic nucleus could be split. Leading scientists dismissed her idea as "ridiculous." It went against what many people thought they knew about atoms at the time. Noddack based her proposal on experiments carried out by the Italian physicist Enrico Fermi. He thought that by bashing atoms with neutrons, he had created a new element, heavier than any other. She suggested instead that he may have split the nucleus and created lighter elements, for which he had not tested.

Noddack's theory was proven correct in 1938. That year, Fermi won a Nobel Prize, in part for these experiments and his erroneous conclusions. Noddack received five Nobel prize nominations in her life, and never won one.

Many years later, Fermi's student and collaborator Emilio Segrè considered why the possibility of fission escaped them, even though they knew of Noddack's paper. "The reason for our blindness is not clear," he wrote.

The person who helped to prove Noddack's theory correct was Lise Meitner. An Austrian Jew, Meitner was among the first women to go to college in Vienna in 1897. She went on to run her own radiation physics lab at the Kaiser Wilhelm Institute in Berlin. Her collaborator of thirty years was the chemist Otto Hahn, one of the scientists who dismissed Noddack's fission proposal.

In 1938, Meitner was forced to flee the Nazis. She went into exile in Sweden. Later that year, she and her nephew Otto Frisch considered some strange findings produced by Hahn and fellow chemist Fritz Strassmann, who were continuing experiments started with Meitner before her exile. The chemists couldn't interpret their results, and Hahn sought Meitner's help. As Meitner and her nephew walked and skied through the wintery Swedish countryside, it became evident to them that they had done exactly what Noddack proposed four years earlier: They had split a uranium nucleus. Meitner and Frisch worked out the math of the energy release, and the numbers bore them out. In their resulting paper, they called the phenomenon "fission."

The news electrified the physics community, and scientists on both sides of the Atlantic rushed to confirm it. Hahn felt threatened by all the action, worried that he might lose credit for the discovery. He hoped that the distinction of discovering fission might give him some protection from the Nazis.

In his correspondence and scientific papers, Hahn began to downplay the role of Frisch and Meitner or leave them out altogether.

Hahn was helped by the wider effort in Nazi Germany to wipe from the record any contributions by former Jewish citizens. Papers coming out of Germany cited Meitner's earlier publications but left her name off as author. Meitner watched in horror from Sweden as the new scientific establishment erased her.

She had a chance to step into the light of history in 1943, when a delegation of scientists invited her to join them in Los Alamos to work on the atomic bomb. Her nephew Frisch was part of the group. Meitner found the suggestion repugnant. "I will have nothing to do with a bomb!" Frisch remembers her saying.

In 1944, the Royal Swedish Academy of Sciences awarded the Nobel Prize in chemistry to Hahn for discovering fission. Nazi Germany forbade its citizens from accepting Nobel Prizes, so Hahn did not receive the award until after the war. Meitner attended the ceremony. The press coverage described her as his "pupil." In his own speeches and interviews, Hahn barely mentioned Meitner; he said not a word about their thirty years of work together.

: :

Noddack, Meitner, the original computer programmers—these are people historians have to some extent registered. I know there are others. Lived unrecorded, traces I won't recover.

At times, the women contributed to their own erasure. Meitner listed Hahn as the lead author on papers for which she had done nearly all the research. This was part of a complex negotiation with a male power structure. It was also a negotiation with the men closest to them.

Stan Ulam's wife, Françoise. She makes explicit that she removed any trace of herself from her husband's memoir, *Adventures of a Mathematician*. In her small "Postscript" to his book, she writes, "*Adventures* is Stan, pure Stan, all Stan—even though he hardly *wrote* a line of it."

He dictated the book to her over a year of travel, and she "edited and assembled the giant jigsaw puzzle into a draft for him to look at and add a few connecting sentences here and there." She served, she adds later, "as his human word processor."

I learned from her daughter Claire that up until her death in 2011, at age ninety-three, Françoise remained deeply opposed to war, and to the bomb that her husband helped invent. She died just as I was beginning work on this book, and I never got to speak to her. "People were constantly wanting her to come and talk to them," Claire told me. "She was always very careful about what she said."

Careful is Old English. Its original meaning, now obsolete, is "full of sadness, in mourning."

I did not feel that right after my mother's death. Instead, a kind of awe flowed through me. I felt closer to my mother, like I knew her better and saw her more clearly. I often sensed her with me.

I felt humbled in those days, brought down to my knees, as the word "humble" implies—from the root for "humus, or earth," close to dirt. All that I thought I knew or believed about life and death and their strict divisions dissolved within me. I didn't question this understanding. It came before language, from my body. I dwelt in that state all summer.

Then something else, a deeper seam—grief—rose up. It happened on a specific day that I remember. I lay still and my body curled inward, head dropping toward the shape—unmuscled, too sharply bent—that took my mother's body as she left. I felt the huge pulse of an infant longing for her physical presence.

Grief is a god, writes the poet Alice Notley, "as in possession." I experience this; at moments it takes me over. In the daily questions I can no longer ask. The silences I cannot break. The divisions I can never cross. Original, utter loss.

GRIEF

"As a school age child, I had a longing for airplanes," writes Jiro Horikoshi.

He came into the world in rural Gunma Prefecture, Japan, in 1903, the year (as he notes in his autobiography) that the Wright brothers flew their first airplane. Growing up during World War I, Horikoshi consumed stories about the world's first air battles among famous European airplanes—Spads, Fokkers, and Sopwiths—and they "excited my young blood," he writes.

"In my sleep I would often dream of flying in an airplane of my own construction, high over fields and rivers, or sometimes close to the ground."

When Horikoshi was in high school, his older brother introduced him to a friend who taught in the new Department of Aeronautics at the University of Tokyo. The science of designing airplanes was new for Japan, and the world, and the small group of students and professors felt themselves stepping into the unknown.

"I studied hard," wrote Horikoshi. "I felt the demand of the times for aeronautical technology would someday rest on my shoulders."

When he graduated, Horikoshi went to work for the Mitsubishi plant in Nagoya, one of Japan's largest military aircraft manufacturers. The company sent him to Europe and the United States to learn their technologies. Japan wanted to establish a domestic aircraft industry, in part because global treaties created after World War I required countries to limit armaments on ships. The focus on air power got around this; Japan could build up military might through airplanes.

Horikoshi worked relentlessly during the day and read aviation magazines at night. He found some time for distractions, though.

He wrote that in the room next to the design section were "many young girls called tracers, who made tracings of simple drawings...Sometimes I used to remark, while having a cup of coffee with my senior engineer or other coworkers, 'Did you see that cute tracer?'"

Trace comes from an Old French word for "mark, or imprint, tracks"—to follow the path of something or someone else. It also has a second meaning, which has to do with domesticated labor: the straps or chains by which an animal pulls a vehicle.

In the 1930s, Mitsubishi selected Horikoshi to lead the design of a new fighter plane. He and his team created a plane unlike any Japan had seen. In contrast to the common biplanes of the day, it had a single wing. Called Type 96, it was faster than any plane in the Japanese military, and better in mock air battles than the European and U.S. models that Japan had purchased for testing.

The Type 96 entered service in August 1937, shortly after war erupted with China. Horikoshi read about his airplane in the newspaper over breakfast one morning in September. The article said his fighter had shot down thirty planes in fifteen minutes over Nanjing, China's capital. "Normally, at home, I never talked about my work," he wrote. "But this was an exception and I showed the papers to the rest of my family."

Horikoshi's airplane stayed in the news throughout the fall. By December 1937, the Japanese military had emptied the skies of Chinese fighters over Nanjing. On December 13, Japanese soldiers marched into the city. Over the next six weeks, they went on a rampage, raping and murdering hundreds of thousands of people.

Historians and politicians debate this number. Many nationalists in Japan deny that a massacre happened. Accounts by witnesses, perpetrators, and survivors speak of burning babies, burying people alive, and men returning to rape the same women over and over.

The reasons for the massacre seem unclear. Japanese soldiers responded to years of propaganda that depicted the Chinese as inferior. They sought vengeance for the brutal battle for Shanghai that preceded Nanjing. The evacuation of China's national government from the city left a void where chaos could unfold.

Conflicting voices fight to claim this history eighty years later. *Remember it like this,* writes the author Simon Han in a beautiful essay on this issue:

> Perhaps in a world that tells us how to feel about our past, a way forward is to ask a different kind of question—not how a scar came to be, but how it hurt. How it continues to.

: :

Horikoshi wrote that after he finished the Type 96, he became ill and had to leave work for some time. Though he does not reveal the nature of his illness, Horikoshi contracted lung infections throughout his life, and would eventually die of pneumonia.

An internet article called "Grief and the Lungs," tells me that traditional Chinese medicine relates lung disease to excessive sadness. A recent paper in the medical journal *Thorax* finds correlations between declines in lung function in older men and chronic anger.

Anger comes from an ancient root that means "tight, constricted, painful." One of its original meanings in English, now obsolete: "distress, suffering, grief."

REMEMBER

My friend Yukiyo Kawano told me she had discovered that her mother's father may have fought in the battle of Nanjing. Yukiyo was born and raised in Hiroshima. She is a third generation *hibakusha*, the Japanese term for survivors of the atomic bomb. Her mother's father lived through the bombing after he returned from the military, and so did her paternal grandmother.

Yukiyo was going through some of her mother's papers and found a government bond worth 10,000 U.S. dollars issued to her grandfather, in gratitude for his achievements during the Sino-Japanese War. Conscripted as a soldier into the Imperial Army, Yukiyo's grandfather served in China from 1937 through 1941. Her grandfather never spoke of this to her; he never spoke of the atomic bomb either.

Remember it like this, writes Simon Han. But what if the charge is not to remember?

When she was in her twenties, Yukiyo married an American Marine stationed at Iwakuni near Hiroshima and moved with him to the U.S. in the year 2000. She felt depressed in Japan, constrained by expectations that she become an acceptable Japanese woman. She wanted to be a visual artist but abhorred what she saw as the emphasis on rote, technical skills in Japan. "This American man seemed like a hero to me, a savior," she said. "I fell completely in love. I romanticized the idea of marriage."

Yukiyo remembered introducing her husband-to-be to her grandfather. "'He's American,' I told him. My grandfather said, 'What happened was...' Inevitable? Justified?" Yukiyo tried to recall this conversation with her grandfather from twenty years ago, and to translate its nuance into English. "He said, 'It was part of the story,'" she told me. "He meant the atomic bomb, that it was part of the continuation of the story of violence that started in China."

Later that night, Yukiyo sent me an email: "I've been trying to find the right translation of what grandpa said that day. It's something in the nature of, 'It is the path (as we know).' There is a feeling of 'it can't be helped,' and thinking back, he probably also referred to me bringing an American man to his house and telling him, 'I am going to America.'"

In the one picture I've seen of Yukiyo's grandfather, he is wearing a dark suit and glasses and looking down at Yukiyo, a toddler in a flowered dress and red shoes, grinning and reaching toward the camera. Yukiyo's grandmother crouches next to her, holding a peeled banana and smiling. Yukiyo's grandfather does not smile. His eyes are shadowed, invisible.

Yukiyo's marriage was unhappy from the start, and she said she doesn't remember much about that time. She had forgotten that conversation with her grandfather until she found the bond, and it came back to her. "I wish I had paid more attention that day!" she said.

We talked of other things: our lives, our art, grieving our mothers. Yukiyo's mother died of cancer at age sixty-five, five years ago. When she speaks of it, her words get more hesitant, with wider spaces of silence. I think she is choosing what she can say and maintain her composure. I sense her grief, a big upwelling, as tightness in the way she holds herself, a flicker across her features she quickly controls.

Yukiyo's grandmother and grandfather also died of cancer, as did her uncle and an aunt by marriage. The children—her mother, uncle, the aunt—all attended Honkawa Elementary, the school closest to ground zero.

Yukiyo says that Japanese speakers don't use the phrase "ground zero." They say *bakushin-chi,* three characters in Kanji that mean:

爆　**blast**

心　**heart**

地　**place**

How

it hurts

Remember

ZERO

After a long absence because of his illness, Jiro Horikoshi returned to Mitsubishi in 1938 to begin work on a new fighter required by the Japanese Imperial Navy. They called for a fighter with range, speed, maneuverability, equipment, and armaments that exceeded any fighter plane in existence.

Horikoshi wrote that fulfilling the Navy's orders meant designing an aircraft so light that it defied common sense. They also had to conserve materials, because Japan had few domestic resources. And they had to compete with Japan's other large manufacturer, Nakajima Aircraft, to win the contract. "The situation was truly merciless," he wrote.

The new plane's impossible requirements grew out of the success of Horikoshi's last plane. To escape the Type 96, the Chinese military withdrew its airfields far inland. The Japanese Navy wanted an aircraft with the range to reach deep into China, and the heavy armaments, maneuverability, and speed to win in air-to-air combat.

The challenge consumed Horikoshi and he noticed little else throughout the spring and summer. "Still thinking that the fields were full of green grass, I was surprised one day, as I sat in the cafeteria, to see the outside green had turned to yellow," he wrote.

Horikoshi and his team of engineers produced more than three thousand drawings over ten months. He pored over every detail for ways to improve the plane's performance. He designed a longer fuselage and larger wings than required to make the plane "a better gun platform" and provide stability from the recoil of the large 20mm canons. He included a wingtip with a slight twist, invisible to the eye, that gave the pilot better control in dogfights.

Most important, Horikoshi focused on saving every gram of weight possible, from large-scale design decisions, like making the wings a single unit from root to tip to avoid heavy attachment fittings, to specifying that one unit of hardware be made of aluminum alloy instead of steel, saving seventy-five grams, or, as Horikoshi notes, about 1/30,000 of the airplane's final weight.

Horikoshi's team worked through the year. By the end of 1938, they had completed the design. Horikoshi took a break. "The New Year's holiday of 1939 was the most peaceful period of time I had known in the last six months," he wrote. "In June of 1937 my son was born, but because I had been so busy I had not been able to spend much time with him. My son enjoyed it very much when I picked him up and raised him high above my head as we stood on the sunny veranda." The vacation lasted three days.

As winter ended and spring came on, the plane began to take shape inside the factory. In April, it flew for the first time, the pilot performing loops and dives above the airfield. Horikoshi wrote:

> I was looking into the sky, forgetting my stiff neck, as I joyously felt the vibration of the air all over my body. The airplane was flying delightfully, wildly and daringly, like a young bird which had finally found its freedom. The trim wing cut sharply through the air and reflected the sunlight every time it turned over. I was almost screaming, "It's beautiful," forgetting for a moment, I was the designer.

The Japanese Imperial Navy officially accepted the new fighter in July 1940, three years after it had first issued the requirements. Per usual practice, they gave the plane a name based on the last two digits of the Japanese year, 2600, dated from the mythical founding of Japan. It should have been the Type 00, but the military shortened the name to Type Zero, the Zero.

: :

The word zero came to English through Italian, or French. Those languages translated it from the Arabic word *sifr*, itself a translation of the Sanskrit word for nothingness, *sunya.* Europe lacked any concept of zero until the twelfth century, when travelers brought it from India. The Sumerians used zero 5,000 years ago. The Mayans had a zero, represented by a man with his head thrown back, a turtle shell, or eye symbol. The circle for zero comes from the Greeks, or the Indians, or possibly from China.

Mathematicians and historians obsess over the murky origins of this number. Most cultures used it as a placeholder, or blank: to distinguish 1 from 10 and 101, for example. The earliest documented use of zero as a number to add, subtract, and multiply with comes from seventh-century India.

Without this number—zero—no modern math is possible, no physics, no Zero airplane, no atomic or thermonuclear bomb. No foam clamshell or plastic umbrella handle.

Yukiyo told me that Japanese speakers use the English word, Zero, for the airplane. Japanese also has its own words for zero: *rei* and *maru,* which means circle. She said that Japanese speakers use *maru* in casual speech for things like addresses and phone numbers, the way English speakers use "O."

The English word circle comes from the ancient Indo-European root *sker* for bend. Used twice, *sker-sker*—circle—it means the thing that bends and comes around again.

RADAR

On September 1, 1939, the day Stan Ulam arrived in New York, the day Hitler's panzers rolled into Poland, a small plant went online in Britain to produce polyethylene. Chemists at the British conglomerate Imperial Chemical Industries had discovered polyethylene six years earlier. They were experimenting with mixtures of different chemicals squeezed at high pressures. They pumped ethylene and benzaldehyde into a compressing tank and left it for the weekend. When they came back, there was no pressure in the tank. They thought it had leaked. When they took it apart to fix it, they found a white wax coating one of the tubes.

The substance seemed so different from polymers known at the time that no one could think of any use for it. But a few of the chemists kept playing around with the mixture in their spare time. Their tests showed that unlike any other plastic in existence, polyethylene was a good insulator for electrical currents. That meant its electrons stayed firmly in place, so that even when high voltages and frequencies passed through, the plastic didn't absorb their energy.

The discovery came just in time for radar, which required high voltages to send radio waves over long distances and receive a reflected signal back. The plastic replaced heavier glass and ceramic and made it possible to create radar light enough for airplanes.

One year later, as the Blitz rained down on London, a group of British scientists and military people traveled to the U.S. to exchange secret scientific information. They demonstrated their new airborne radar on an American PBY, the type of plane flown by Christman. The Navy people liked it so much they ordered seven thousand, and PBYs became the first Navy planes to carry radar.

Radar did not do much to improve accuracy, but it did increase war's brutality. The most advanced radars of the time only gave accuracy to within a few miles, the size of most city centers. But unlike the Norden bombsight that Christman used, radar did not require seeing the target. In Europe, covered almost constantly with clouds during fall and winter, radar made continuous attacks possible. It cemented the strategy of terror bombing cities, an approach for which the atomic bomb turned out to be the ultimate weapon.

The doctrine of targeting cities evolved in part because of technology, in part because of each side's increasing desire for vengeance. In July 1940, German airplanes started bombing English airfields, docks, and railyards to open the way for a land invasion. Germany had four times as many planes as Britain, so the Reich expected an easy victory. But Britain had set up a radar network along its coast called Chain Home. It gave the Royal Air Force enough early warning to shoot down German planes.

In the midst of this, on August 24, German pilots dropped bombs on London by mistake. Churchill went into a rage and ordered retaliatory strikes on Berlin. In response, Hitler dropped more bombs on London.

Britain's radar system kept the Nazis from knocking out the country's defenses so they could invade. Instead, in September, the Germans settled into bombing cities. They had used a similar strategy in the Spanish Civil War, bombing Guernica and other population centers to terrorize people and break their will for fighting. The Blitz lasted nine months, until spring 1941.

In February 1942, Britain adopted the terror bombing of cities as official policy. The U.S., at first opposed, signed on a year later when Roosevelt met Churchill at Casablanca and agreed to an Allied bombing strategy focused on "undermining the morale of the German people." The campaign reached its pinnacle under the head of British Bomber Command, Arthur "Butch" Harris, with the firebombing of Dresden on Valentine's Day 1945, which leveled the city and killed tens of thousands, horrifying Britain's own people.

But Butch Harris had nothing on Curtis "the Demon" LeMay, head of the U.S. bombing operations in the Pacific. A month after the bombing of Dresden, LeMay launched the most violent air strike yet against the people of Tokyo. Radar worked well for bombing Japan because most of its cities sit on the coast, making them easy to distinguish from the water.

On the night of March 9, more than three hundred U.S. bombers dropped two thousand tons of bombs on downtown Tokyo. The *aka-kaze*—red wind—off the plain of Tokyo fanned a tidal wave of flame. More than one hundred thousand people died, the deadliest bombing in history.

And LeMay was just getting started. By the time the U.S. had atomic bombs to drop, the military had to pick secondary targets because LeMay had already destroyed all of Japan's largest cities. LeMay famously said that if he had lost the war, he would have been tried as a war criminal.

The pilot who shot Elwyn Christman out of the sky that day in 1941 flew one of Horikoshi's Zero airplanes. Every fighter plane engaged in the attacks on Pearl Harbor and the Philippines came from the mind of Horikoshi and his design team, either Type 96s or Zeros.

The Zero took the Allies completely by surprise. Claire Chennault, a retired U.S. fighter pilot and Army officer, advised China on its air strategy in the Sino-Japanese war. He repeatedly tried to warn U.S., British, and Australian officials about the Zero. They ignored his reports. No one believed the Japanese could create such an advanced machine. They concluded the defeats in China must have resulted from a lack of skill among Chinese pilots. To Western powers, the Zero remained a blank, invisible.

When it appeared in the skies over the Pacific, the Zero quickly gained the status of myth. The fighter seemed everywhere at once; it brought down some eighty percent of Allied aircraft destroyed in the first months of the war. At first, many believed it must be a design stolen from the West, but the U.S. and Europe had nothing like it. U.S. pilots soon found their best defense against the Zero: flee from it.

: :

Horikoshi was at home with another lung illness when he heard the news of Pearl Harbor on the radio. He tried to go back to work, but severe pains in his back and chest drove him to bed.

He calls World War II "the days of the frightening dreams."

At first, he rejoiced in the news of his airplane's victories. But he also felt uneasy from the moment he learned about the Pearl Harbor attack. Horikoshi knew that Japan could not win a prolonged war against the U.S., with its vast reserves of raw materials and industrial capacity.

When Horikoshi returned to work in early 1942, supplies already were short. "Steam for heating the office had been cut off and the quality of even our paper was visibly lower," he wrote. "In the center of the design room a wood stove was burning, but in the assembly plant, where rivet guns were hammering away, there was no heat at all and the airplanes were being assembled on the cold concrete floor."

The lack of flat, unfarmed land in Japan meant that the airfield was some thirty miles away from the Mitsubishi assembly plant. Teams of oxen moved every Zero airplane built—more than 100 per month at the height of the war—over the rough gravel roads between plant and airfield. The animals were the slowest form of transportation (it took 24 hours), but the most stable. Bumping trucks might damage the airframes. Drivers plied the oxen with food and beer to keep them going, but by the end of the war, all the animals had died from exhaustion. The military switched to horses, and most of those died also.

::

In June 1942, the tide of the war turned when the U.S. military forced Japan to retreat from Midway Island. Midway was supposed to be the battle that led the U.S. to concede defeat to Japan. But the U.S. broke the Japanese Navy's coding system, learned of the attack, and planned a counterassault. The U.S. Navy's radar system, a technology that Japan lacked, let the military detect approaching Japanese ships.

Horikoshi writes that the Japanese press reported Midway as a victory, and the military attempted to hide and obscure the defeats that followed. But Horikoshi could guess the truth from the increasingly urgent demands for airplanes and pilots. People donated metal for aircraft, and plant employees searched out pine roots and grew castor oil plants for fuel and lubricants.

On December 7, 1944, a strong earthquake hit, killing nearly a thousand people in the Nagoya area and cracking the factory floor, knocking apart the jigs for assembling airplanes. Ten days later, U.S. B-29s bombed the plant. Horikoshi remembered:

> I took shelter in a vacant lot just outside the buildings when I heard the siren's piercing cry. As I lay face down in the ditch, I looked up out of the corner of my eyes toward the roaring sounds and saw a column of more than ten beautiful B-29s leisurely approaching...I tensed my shoulders at the sharp noise of the bombs falling through the air.

The raid killed and injured more than four hundred workers. The historian Akira Yoshimura described the aftermath: "Limbs were to be found all over the works. Some were even caught in the beams or electric wires...As night fell, the last corpse was collected. But the scraping of flesh attached to unfinished wings, or pillars, and walls went on."

In November of 1944, the Japanese press unveiled the Zero to the public for the first time, hoping to build morale, and announced "sacrifice attacks" by kamikaze pilots.

Most airplanes used in the thousands of kamikaze attacks were Zeros. "Airplanes became a kind of bomb with a youth aboard," writes Yoshimura. "The Pacific Ocean became a grand suicidal place for youth who wanted to save their country...To the [Imperial Japanese] forces, the pilot was no longer human but merely a device for controlling the bomb."

Asahi, the main newspaper in Japan, asked Horikoshi for an essay in praise of the kamikazes, but he could not bring himself to write anything. He finally handed over the essay in early 1945. In his book, published in 1970, Horikoshi tries to put what he wrote in a historical context that registers the horror and opposition he felt then but couldn't publicly speak. "I strongly wished to include at least a paragraph of protest," he says, and quotes the following from his essay: "We have reached the limits of human intelligence and have selfishly tried all kinds of methods to make effective adjustments to our limited human and material resources so that new arms could emerge."

: :

Horikoshi barely speaks about himself or his family in his book. But an earlier book called *Zero!*, first published in 1956, purports to contain entries from Horikoshi's diaries late in the war. The book was translated by Martin Caidin, a prolific U.S. novelist and aviation historian. In his journal, Horikoshi writes that he became ill with pleurisy and had to leave work from December 1944 until the end of July 1945. He records hearing about the firebombing of Tokyo and writes that the U.S. firebombed his own city, Nagoya, the next day. He had to be helped into the air raid shelter as screeching firebombs rained down, destroying much of the city.

In his later book, Horikoshi skips over his illness and writes dispassionately about the firebombing. He records nothing about the atomic bombs, but he recalls hearing the Emperor's announcement of surrender when he was home from work for lunch at noon on August 15, 1945. "I thought, 'This finishes the job to which I have devoted one half of my life.'"

Horikoshi died of pneumonia in 1982. A profile of him in the *Los Angeles Times* mentions that he had five children, "none of whom became involved in the aircraft industry."

WIND

Because Horikoshi revealed almost nothing about his life, Hayao Miyazaki made it up. In 2013, the celebrated Japanese filmmaker released *The Wind Rises*, an animated feature about Horikoshi that Miyazaki claimed would be his last movie. It was the top grossing film of the year in Japan.

The movie features lavishly illustrated, bucolic scenes of prewar Japan, and gorgeous dream sequences of Horikoshi flying fantastical airplanes that resemble birds or sea creatures. The Italian airplane designer Giovanni Battista Caproni appears in his visions as a figure to guide and inspire.

Horikoshi is depicted as dreamy and bespectacled, imbued with heroism. As a boy, he protects another student from bullies. Later, as he rides a crowded train to Tokyo, a powerful earthquake strikes: the Great Kantō earthquake of 1923. Horikoshi helps a young girl and her governess, who has broken her leg. He carries the governess on his back through the burning city and reunites the two with the girl's family.

They do not meet again until years later, when Horikoshi encounters the girl, now grown. Her name is Naoko Satomi. She says that they looked for him for years. "You see," she tells him, "you were our knight in shining armor."

They court one another, flirting and playing, but all is not well. Naoko is sick—Miyazaki gives her the lung disease instead of Horikoshi. Even though she is gravely ill with tuberculosis, the two decide to marry. Horikoshi drags his table to her bedside and holds her hand as he designs the Zero.

Naoko is the passively tragic foil for Horikoshi's strength and heroism. But, for a hero, he also is strangely passive. Throughout the movie, the plane designers are depicted as lovers of knowledge and beauty swept along by impersonal forces. Caproni tells him, "Humanity has always dreamed of flying, but the dream is cursed. My aircraft are destined to become tools for slaughter and destruction."

Even when Caproni shows Horikoshi the ultimate fate of his creations—kamikaze attacks—Horikoshi shouts, "I just want to design beautiful airplanes!"

Miyazaki said in an interview that this quote inspired him to write the movie. It is full of personal echoes: Like Horikoshi, Miyazaki dreamed of airplanes as a boy, and drew them obsessively. His film studio, Ghibli, is named after a warplane designed by Caproni.

The filmmaker's father and uncle owned a factory that made parts for the Zero during the war, and Miyazaki grew up in privilege. He remembers the U.S. firebombing his city, Utsunomiya, when he was four years old. Unlike many other residents, Miyazaki's family had a company truck to escape. He remembers a woman holding a child and running toward the truck, but the adults did not stop.

: :

Miyazaki uses wind in the movie as a metaphor for fate, vast forces shaping the lives of the characters. The movie opens with an epigraph by the early twentieth century French poet Paul Valéry: "The wind is rising!...We must try to live!" It comes from his poem, "The Graveyard by the Sea."

When Horikoshi and Naoko meet each other on the train, she quotes the first half of the line to him in French, and he completes it. Each time they meet, a gust of wind arises. At the end of the movie, as Horikoshi watches his Zero soar through the sky, his moment of triumph, the wind comes up and he knows that Naoko has died.

In Valéry's poem, the wind seems less like fate, and more like the ceaseless, transforming power of life. The speaker glories in the sparkling sea, and addresses it as a mythical creature, an ouroboros:

> Yes, mighty sea with such wild frenzies gifted...
>
> Creature supreme, drunk on your own blue flesh,
>
> Who in a tumult like the deepest hush
>
> Bite at your sequin-glittering tail—yes, listen!
>
> The wind is rising...We must try to live!

: :

In the movie's final moments, Caproni and Horikoshi watch Zero airplanes fly off into the sunset like beautiful birds, joining a flock of thousands more. "Not a single one returned," says Horikoshi, a reference to the kamikaze pilots. But Caproni's response puts the focus on machines, not people. He says, "There was nothing to return to. Airplanes are beautiful, cursed dreams, waiting for the sky to swallow them up."

This poeticism, the movie's emphasis on fantasy and fate, brackets out the lives destroyed by forced labor, war, and imperialism. It portrays these as impersonal forces, not human creations driven by those in power and maintained through countless individual choices by people all over the world like Horikoshi, engaged in trying to live.

TO LIVE

In 2017, Miyazaki came out of retirement to work on a new film, *How Do You Live?* The press describes the movie as a farewell to his grandson. His producer, Toshio Suzuki, told the Japanese media, "It's his way of saying, 'Grandpa is moving on to the next world, but he's leaving behind this film.'"

The film takes its name from a book published in 1937, the year Japan went to war against China. It appeared as part of a series of books for young adults that supported liberalism and progressive ideals. With Japan's entry into war, the Imperial government took a different tack. It promoted nationalism, loyalty, the power of the Emperor, and service to the state.

After World War II, U.S. forces occupying Japan changed course again. They issued directives to curb Japan's militarism and oversaw the drafting of a new constitution that limited the power of the Emperor and renounced war.

Conservatives now in power in Japan seek to revive wartime values. Japan's longest-serving prime minister, Shinzō Abe, who stepped down in 2020 because of illness, is the grandson of Nobusuke Kishi, brutal overlord of Japan's colony in Manchuria during the 1930s, where officials forced Chinese people into slavery to produce coal, steel, and weapons. Kishi returned to Japan and became Minister of Commerce and Industry during the war. After Japan's surrender, the occupying forces imprisoned him as a war criminal. The U.S. freed Kishi and others in 1948 because they wanted conservative leaders in Japan to form a bulwark against China's communism. Kishi served as prime minister from 1957 to 1960.

After he was elected prime minister in 2012, Abe went to the tomb of his grandfather and pledged to fulfill Kishi's ambition to "recover the true independence of Japan." The ultimate goal: to remove the constitution's renunciation of war. His successors still hold that ambition.

Remember—

how it hurt. How it continues to.

: :

My mother did not like my collaborations with Yukiyo. She carried a bodily memory of her father's war trauma, and vehemently reminded me of Japan's wartime brutality. She was a toddler when her father went to war; her ferocity surprised me.

After she died, I suddenly felt like I could ask my aunt again about my grandfather. We had this conversation over Facebook Messenger. Here's what she wrote:

> Dad drove the Higgins boats that dropped off troops to the beaches...Mom told me that after he came home, he would start to stab her at night while they were asleep. A friend of his from the war came to visit, and he told mom that dad's ship was sunk by kamikazes while he was dropping off guys to the beach. He didn't have a ship to go back to so he had to go on the beach and fight. All he had was a knife. He killed at least one Japanese man. He kept it secret for years. I came home from college one year and my Lutheran pastor had given a sermon that killing, even in war, was a sin. So I brought up the subject and said that I disagreed with the pastor...Then dad told me that he killed a man in the war and it was very hard for him to live with that, but it was him or the other guy.

I admit that I am fascinated by this story. How my grandfather—the gentle Swedish giant I knew, deeply religious, who loved to sing—could kill another person with the intimate violence that a knife requires. How he lived after, constructing a life around this submerged fact.

How it hurt. How it continues to.

: :

> In March 1945, 31-year-old Tsunematsu Tanaka sent his two little children, Makio and Shigeri, and his wife, Mikie, to a little country town, Miyoshi, located 30 miles northeast of Hiroshima. For himself, he rented a room in a small family house near the city. It was an easy distance for him to go to his work, a local electric company located in downtown Hiroshima. On August 6, 1945, he arrived at work early at around 8:00 am. He decided to go downstairs to a laundry tub in the basement—not knowing that his life was at stake as he was stepping down the stairs—and was washing the uniform that he wore the night before when he was on duty patrolling the city during the nightly air-raid alarm. It was a hot sunny morning. The sun was blazing and the temperature was quickly rising. It was a chance to cool down before he would start working at 8:30. At exactly 8:15 am, the blast occurred, followed by the intense white light. Broken glass splinters pierced him from behind as he was kneeling down. It is believed that he was knocked unconscious for several minutes. He awoke, saw the fire starting from the upper floor, and pulled himself out before the fire wiped out the whole building. He went south and took a ferry to the little island Noinoshima to seek medical care. The medical facilities near the city were filled with the unrecognizable blackened flesh covered with yellow pus of those alive and dead. He sensed that he would die there if he stayed. Early next morning, Tsunematsu walked seven miles north to Hesaka Station and got on a train to Miyoshi.

Yukiyo found her maternal grandfather's atomic bomb testimony after he died in 2008 at age ninety-four. It had been recorded by a civil servant, part of the government's outreach program to hibakusha, months before he died. She says she asked him to tell her the story many years ago. He said, "No point. You won't be able to understand."

: :

Yukiyo constructs diaphanous life-size sculptures of the Hiroshima and Nagasaki bombs out of kimonos made by her paternal grandmother, also an atomic bomb survivor. Yukiyo sews the kimono fabric together with strands of her hair, weaving her DNA—and the damage it may carry—into the silk sewn together by her grandmother.

Yukiyo has written that she was inspired to make the sculptures when she read the hibakusha poet Sadako Kurihara. "Ms. Kurihara declares that an epigraph in the Hiroshima Peace Memorial Park that pleads, 'Please rest in peace / for we shall not repeat the mistake,' should have said instead:

O deceased

do not rest in peace.

. . . .

You must perturb and awaken

the avaricious living dead

The first evil

may have been a mistake

but the second is a treachery

Faithfulness to the dead

we shall not forget.

"I decided that I wanted to try making Little Boy, the A-bomb dropped on Hiroshima, out of the disassembled kimono," writes

Yukiyo. "Three days later, on March 11, 2011, I heard the news of the Tohoku region earthquake followed by a twenty-meter tsunami that hit and paralyzed the Daiichi nuclear plant in Fukushima... The stitch [of my hair became] very fine as I prayed for the over 18,000 dead and missing, while I listened to the special internet program that broadcast for a month following the earthquake."

: :

I'm considering why I felt suddenly free after my mother died to ask about my grandfather. I think I understood instinctively that she would not want to face this difficult story about her father. When she was dying, she didn't want to face other people's grief, either. She refused visits from most people, including her closest friends; including my aunt, her only sibling.

I wrote her a letter as she was dying, trying to capture some of what she'd meant to me. It was too hard for her to read it, so I knelt beside her and read it to her. Her face looked pained, grief-filled, and she did cry, she cried out, a little. It's the only time I saw tears from her about her death. I put something in the letter I didn't fully comprehend at the time. I wrote, "You are my gift and my legacy, and I'm yours."

How could my mother be my legacy, something I leave for the future? Now it makes perfect sense to me. By remembering, by remembering across generations and without refusal our pained entanglements, and by being responsible, answerable to that pain, we can carry each other into a past that might make a living future.

THE THINKER

Claire Ulam Weiner did not want me to record her. She agreed to meet me at a cafe in Santa Fe, all the while protesting that she is no expert on her father and not the right person to discuss his work. She was kind (she bought me a hot chocolate) but firm. She opened our conversation by firing off a list of people to consult and books to read. I had a hard time admitting what I longed for: just a sense of him as a living, breathing person. I certainly didn't lead with the question about what his eyes were like.

Claire's eyes are dark, so dark the pupil and iris merge. They spark.

She claimed a typical immigrant childhood: She wanted to be more American than her parents. She threw herself into school life and social activities. She didn't take much notice of what her parents were doing. "I wasn't interested in science," she told me.

She also knows little about her father's Polish past. "He really shut that off," she said. She didn't learn she had an aunt and a baby cousin killed by the Nazis until she was in high school. She referred me to another cousin: Alex, the son of Ulam's younger brother, Adam, who became a renowned historian of the Soviet Union. Alex, a journalist, had just visited Lviv in Ukraine (formerly Lwów) and written about it for *The New York Times*. "He's the one to ask about family," she said.

She was neither curt nor impatient. I sensed she would never be rude enough to betray impatience. I got the feeling, though, that her mind moves very fast, and that she was eager to be on to the next thing, not inclined to divulge much to me. It is a bit of a burden, this meeting with people who have an interest in her father; a burden her mother, Françoise, used to carry.

I ventured to ask about another cousin, the one who made the sketch of her father on the cover of his memoirs. She didn't know anything about him, but she knew the drawing. "The eyes!" she said.

"The eyes," I agreed. And then I could ask her.

She shrugged. "He had green eyes, like yours," she told me. But I know his eyes were not like mine. Mine are hazel, brown in certain light. Too dark. In my mind, his eyes are luminous, blue-green, oceanic. I didn't want to contradict her, though, so we sat in silence.

She moved for her bag. Fearing she was on the verge of leaving, I mentioned Françoise's recollection of Ulam sitting in the living room and staring out the window, thinking. "Oh, he was always like that," she said, leaning toward me. "He was always just sitting, thinking. Or he would lie on the couch, thinking. Or he would play solitaire and think. In fact, *The New Yorker* one year when I was tiny published a little squib about a young girl in Los Alamos who said her father never played ball or did anything but think. That was me."

I have searched old issues of *The New Yorker* and failed to find this episode. But it does appear in the book *From Cardinals to Chaos,* a collection of essays dedicated to Ulam. The title of the book refers to Ulam's discoveries in mathematics. His exploration of large cardinal numbers opened up new orders of infinity for mathematicians to explore.

As for chaos, work by Ulam and others formed the foundations of chaos theory: that within apparently random systems lie underlying patterns and interconnections. Chaos is not chaos after all. And here lies another occluded woman.

In 2008, *Physics Today* published an article, "Fermi, Pasta, Ulam, and a mysterious lady." The word "lady" comes from Old English. It meant "mistress of a house, wife of a lord," but combines two words that meant "loaf" and "kneader," the maker of bread. In this case, the "lady" was a computer; her dough was numbers.

In the 1950s, after the hydrogen bomb, Ulam returned to ideas of theoretical interest in math. He, Enrico Fermi, and another mathematician named John Pasta wanted to explore how a vibrating string propagates energy. To simulate this, they relied on the MANIAC computer, the successor to the ENIAC, and on Mary Tsingou, a mathematician and programmer who created the intricate algorithm and coded in the program.

The results at first showed the predicted random patterns, but at one point, someone forgot to turn off the computer. When the simulation ran for long enough, it showed that the energy followed a complex, repeating pattern. This became known as the Fermi-Pasta-Ulam paradox.

The report about this discovery, published in 1955, lists Fermi, Pasta, and Ulam as authors, and credits Tsingou for "work done," with a footnote thanking "Miss Mary Tsingou" for running the computations. The author of the 2008 *Physics Today* article, Thierry Dauxois, notes that early computer programmers often received credit as coauthors. The fact that Tsingou did not actually write the report might account for her second-tier reference. But as Dauxois points out, Fermi didn't write the report either; he had already died of cancer.

The mystery of Tsingou's identity deepened when she married and took her husband's name, Menzel. In 1972, she repeated the simulation with the physicist James Tuck and appears as the second author on that report under her married name, leaving her original contribution unacknowledged. Dauxois, a physicist, determined to give Mary Tsingou-Menzel proper credit for this now-famous paradox in science, by renaming it the Fermi-Pasta-Ulam-Tsingou problem.

The attention embarrassed Menzel, who remained at the lab for forty years and still lives in Los Alamos. "I was not a mysterious lady," she told the local paper. "We were all young people. The Korean War was on, so they hired women mathematicians...They treated me as an equal."

Menzel grew up in Milwaukee, Wisconsin, the daughter of a Greek family that immigrated to the U.S. from Bulgaria. She earned a degree in mathematics and education, as did her older sister, who worked as a programmer on another early computer. Despite her claim to have been treated as an equal, in an oral history conducted in 2002 Menzel said that she always faced obstacles to getting promotions and equal pay. "They told us we would be doing the same work as males but we would get less pay and we accepted this," she said in the *Los Alamos Monitor*.

Menzel worked with Ulam for years at Los Alamos, exploring theoretical problems he raised. They often did this work on nights and weekends; her main work was simulating nuclear explosions.

The oral history interviewer asked Menzel if she had any mentors or role models at Los Alamos. She said, "Well, some of these guys were pretty nice to me. When I went and got my Masters and came back, Stan Ulam tried to get me a 25 dollar-a-month raise! [Laughs.] Things like that. But as I said, with women, we were always second citizens."

: :

The drawing of Ulam in the book *From Cardinals to Chaos* is a montage of images, with Ulam in the center. Three white lines converge in his left eye, the good one. The top line connects to a drawing of ovals bisected by a triangle. The bottom line trails off into space. The center line extends from a sketch of Omega Bridge, soaring off its graceful arch. Ulam floats beside it. The bridge connects the town of Los Alamos to the nuclear lab across the canyon.

I can find no information about who named the bridge Omega, the last letter of the Greek alphabet, a mathematical symbol, and the title God claims, according to John, in the *Book of Revelation*: "I am the Alpha and the Omega, the first and last, the beginning and the end."

My dad wrote this story to me in an email about his first trip to Los Alamos as a twenty-seven-year-old, freshly minted physicist from the University of Iowa:

> When I came to interview in May of 1970, whoever routed me from Iowa City to Los Alamos must have drawn straight lines on a map since it took all day and four separate flight segments to travel a thousand miles. When I arrived, it was dark and I was travel weary. I woke early the following morning and walked outside the Hilltop House [hotel] and to the edge of the canyon. The sun was shining brightly. It was a beautiful morning. The bridge was illuminated, so much so that it seemed to be floating in space, an almost perfect capital Greek omega.

My father is the only member of the family ever to get a Ph.D. He wrote an essay when he was twelve years old, in 1955, that said he wanted to be a physicist. At least that's the family story. He told me recently he doesn't remember this, only that he always knew he wanted to do physics. "There was never any question," he said. I asked him why. At first, he said he didn't know.

Then he said his mother bought a set of encyclopedias from a traveling salesman when he was about eight, and he read the science and it fascinated him. When he was in high school, physics was a hot field, full of discoveries about the mysteries of matter. His Ph.D. thesis involved high energy particle physics, matter at the subatomic scale.

He met my mother in eighth grade, about the time of the essay. She said that he wore pink pants to school one day, and she thought he was cute. She worked as a dental hygienist to put him through school. They moved to Los Alamos in the summer of 1970, when she was pregnant with me. She cried as they drove through the brown desert up seven thousand feet to the town, clinging to mesas spread like fingers from the Jemez Mountains. She thought her nausea and dizziness came from altitude sickness.

My dad spent most of his career at Los Alamos in nonproliferation, working to stop the spread of nuclear weapons. One of his titles during his long tenure: Director of Threat Reduction.

For most of my growing up, my mother refused to cross Omega Bridge. She was afraid of it.

: :

The collection of essays written in tribute to Ulam originally appeared in *Los Alamos Science,* the journal produced by the laboratory. In the Los Alamos version, Claire appears on the very first page. When Cambridge University republished it as a book, the editors moved her to the very end. Alpha and Omega. First and last. She appears aged five or six, with a pageboy haircut and a Mona Lisa smile below those dark eyes. The text below says:

> One day when Little Claire Ulam was watching some children playing ball with their father, a friend asked whether her father ever played like that with her. The answer was an emphatic, "No! No! All my father does is think, think, think! Nothing but think!"

SEED

The basic unit of existence is not the individual, but the relationship. The physicist Karen Barad concludes this, based on the principles of quantum physics. She writes:

> To be entangled is not simply to be intertwined with another, as in the joining of separate entities, but to lack an independent, self-contained existence...Individuals do not preexist their interactions; rather, individuals emerge through and as part of their entangled intra-relating.

She uses "intra-relating" instead of "inter-relating" to denote this. All that exists is merging and overlapping phenomena. The implications for Barad are not only scientific. Or, rather, she recognizes no distinction between what is science and what is ethics, what is knowledge and what is existence.

I am of this: my mother, in cell and bone and breath. How could I not be? One nervous system takes shape enfolded within another. I am plastic blister pack, calculus, computer. I am nuclear weapon, and car part, and war. I am—deeply—of my father.

I am of Jen, also, this "we" out of twenty years twined together. But something happened. Our entanglement started to fray and falter. Loss. Jen and I began to split.

And then I was—what? Disoriented to everything. Vertigo seized me. I could only hang onto the bed and crawl some mornings. Some days I could only shake. Some days I felt trapped in a burning building with an alarm going off, unable to escape my own skin. I clawed my way through moment to moment, incapable of lifting my head to see anything past that.

But Jen and I persisted. We hung on. Sometimes it seemed almost unwilled, a deep allegiance to—maybe just being itself as we entangledly knew it.

In time, we came through. We became something new that we could recognize. A new form of family, not paired, but twined still together. A "we" still alive with the blood and the love we have fed it.

And a new love came to my life. It happened like this. Or, this tells it well: In those years, I traveled weekly to a wise, compassionate counselor, one of a handful of female teachers and guides I drew to me at that time. She had her office in a beautiful old building in far Northwest Portland. A gnarled birch tree towered out front. As I trudged up the brown carpeted stairs to her office each week, I noticed that others before me had trailed in birch seeds on their feet, and the seeds were shaped like birds. This seemed lovely to me, a glimmer of hope, some possibility. I collected a few. One I kept nearby on my table. One I swallowed one morning in a glass of water. I don't know why, some instinct. I wanted the bird entangled with me.

Months later, I met Cobalt. Slowly I enfolded into family with them and their daughter, Aster. Cobalt's house in North Portland sits between the freeway, the train tracks, and the scrap metal recyclers and auto wreckers along the Columbia Slough. Their wooded yard feels like a sanctuary, their small house shaded by a gathering of birch trees that releases a deluge of bird shaped seeds. I started to find them everywhere: in my shoes, on my clothes, in bed, in the rugs, on the dogs. Birch seeds rained down on me.

HERE

In fall 2018, I went back to Hawai'i to visit Kamilo and to see Noni, the artist who makes objects from the plastic she finds on the beach. I used the miles saved up from flying to New Mexico over and over during my mother's illness, death, and funeral. Cobalt came with me (2,500 pounds of pollution into the atmosphere). They introduced me to Bill Gilmartin, co-founder of the Hawai'i Wildlife Fund, which runs the beach clean-ups at Kamilo. Bill is a scientist who spent years studying endangered Hawaiian monk seals.

We met Bill early in the morning at his house in Volcano, tucked into the rainforest that drapes the eastern flank of Kīlauea. He was tall and somewhat stooped, with a long, craggy face and grey ponytail. Mattie Mae was there too, the Hawaiian-born clean-up coordinator for the Fund. She was a blur of motion with her little dog Pua, loading the truck with supplies for the trip. They had organized a volunteer beach clean-up around our visit.

We pulled up at Kamilo and I was again struck by its stark beauty, turquoise ocean waves frothing over black lava rock. We headed toward the beach with its heaps of driftwood and sun-bleached plastic. Mattie wore her long hair in a braid, and covered herself with a hat, glasses, a cowl, and long sleeves to protect from the sun. She handed out buckets and large plastic sacks with instructions to pick up the big pieces only, for efficiency's sake. "Big means a foot or larger!" she shouted after us as we fanned out across the beach.

It was hot and still at Kamilo this time, with barely any wind. I picked my way across the sharp lava, filled a bucket with plastic, brought it to one of the four waiting pickup trucks, then returned to fill it again. Sweat stung my eyes and trickled between my shoulder blades. A feeling of futility crept into me; I left more plastic behind than I carried.

Most of the volunteers were women or older couples. Many were residents, friends of Bill and Mattie. One couple had just retired to Hawai'i from Missouri. A few, like us, were visitors. One was an artist, a Big Island resident, who makes jewelry out of nurdles, the tiny pellets that emerge from plastic factories and get shipped around the world to be melted down and molded into products. She sat on the beach with a test tube and tweezers, collecting nurdles the size of lentils.

I remember this moment: hauling a huge tangle of fishing net off the lava. With no Unimog this time, it took all of us, the middle-aged, the elders, and the young, counting three and shoving, dragging it across the sharp rock and wrestling it up the rickety wooden ramp into the truck. Some of us fell down, scraping shins and knuckles. Some of us fell down and needed help getting back up. Cobalt suggested we grunt. So we grunted, straining and stopping to take breaths, on command of our leader, Mattie Mae, calling: "Ok, one, two, three!"

Tears welled up behind my sunglasses as I grunted and strained with the others. I realized how few times in my life I'd been part of a communal effort like this and felt the power, and the frailty, of our entanglements. Still raw with grief for my mother, raw with loss in general, I cried: for our smallness, our laboring, breathing bodies, etched with injury, illness, mortality. I cried for the hopelessness of the work—a single tangle of net on a single beach among countless others—and for the beauty of the commitment to it anyway, the seriousness of the struggle.

We ate lunch, crouched in what little shade we could find by the trucks, and drove back, hot and exhausted. Bill slowed for us to see a flame-orange wiliwili tree in the distance that he said he'd never seen bloom so profusely. Days before we arrived, Hurricane Lane—the wettest tropical cyclone on record in Hawai'i—dropped fifty-eight inches of water, flushing the normally dry, lava-covered land with color.

On the drive back, Bill told us about finding a fishing net with an arm bone caught in it. "Female, we think, from the size," he said. At first, I thought he was talking about a human, but he meant a monk seal. I wrote down what he said, "Her flipper caught in the net and she had to try and live, dragging that around with her."

We drove in a convoy to the county dump and backed the trucks up to the edge of the garbage pit. We formed a human chain, handing bags and buckets from one to the other, tipping them down a concrete ramp. It took no time at all to dump the plastic we had gathered. Mattie climbed up in the empty truck bed with a broom, sweeping tiny bits of plastic and sand into a dustpan Bill held for her. I took a video of this: a snow of plastic particles missing the pan and falling through the cracks in the tailgate, leaving a colorful confetti behind in the dirt.

I kept two things from that trip. One is a brown octopus trap with Korean writing, torn by large bite marks from some animal, maybe trying to get the octopus out. One is a liter water bottle, tall and thin, with Arabic lettering. In English, it says "frida Bottled Drinking Water" with a white heart on a blue label. It has a tiny round smiley-face sticker that says, in Arabic and English, "Smile you in UAE."

But I was not in the United Arab Emirates. I was on the other side of the globe in a completely different ocean basin. Lots of Pacific Rim detritus washes up on Kamilo, with lettering from the U.S., China, Japan, Korea. Arabic is rare. No one had seen it before. The date on the bottle is June 1, 2018, so it was not at sea very long. It is hardly worn or weathered. It would still work well to hold water. Mattie suggested it got tossed off a passing cargo ship.

I look at this bottle a lot. I consider the person who held it and drank from it, whose life I can't really imagine, from a place I've never seen. I learn that in the United Arab Emirates all drinking water is desalinated seawater. Some people sucked water from the ocean, heated it, and collected the salt-free vapor. Some others made containers from fossilized sea creatures to hold the de-salted seawater. Then a person opened this bottle and drank from it.

The person swallowed the water, along with the plastic and chemicals it carried, then tossed the bottle out into the ocean. Or maybe it blew or washed off the ship. Ocean tides and currents carried the bottle to Kamilo, where it found my hands. I brought it 2,600 miles on an airplane across the Pacific to Oregon.

By zooming way out on Google maps, I can see the whole globe. I rotate it back and forth, from the United Arab Emirates on the Persian Gulf, through the Arabian Sea and the Indian Ocean, across Southeast Asia and Australia, to the specks of Hawai'i in the center of the North Pacific, and further to the coast of North America. Doing this makes clear: It's a single ocean we share.

Smile. You are here.

HAUNT

We stayed at Noni and Ron's house in Volcano after the trip to Kamilo. Noni greeted me with the same gruff kindness as before and took us inside their modest house, lined with windows and surrounded by dense green ferns and forest.

After dinner, Noni and I spent a long time looking through her plastic. I took photos with my camera flash turned on in the dim light of her house. Noni organizes her items largely by type: buckets of toothbrushes, a shelf of toy soldiers, a wire basket full of combs, boxes of lighters.

Among the stranger things in her collection: dozens of yellow plastic jar lids that say "Creap"—a popular Japanese powdered creamer—and three tubs of curved umbrella handles, which she collected over fifteen years. She has wooden printer's trays and shelves full of toys. One is mostly heads or large-headed creatures, their round shapes echoing from shelf to shelf. Toy tops fill another. One has game pieces: dominoes, chess, checkers, and octagonal tiles with Chinese characters. A set of five wooden shelves holds animals, some so abraded by sand and water they are only lumps suggestive of animal shapes—two headless brontosauruses, a headless chicken, a legless blue hippo, and what might have once been a lamb.

Some people collect ocean detritus for treasure, some for science, but Noni looks for—not aesthetics, that word is too flat. She is an artist from a family of artists and she has a highly refined visual sense, but that's not exactly what she's after. She responds, I think, to the haunting in these plastic objects, the crackle or spark that travels across—trace of a living touch.

The word "haunt" has an unknown origin. It comes, perhaps, from an Old Norse word that means "to bring home."

Noni said the collection makes her happy. She said, "I like to have another crazy person around to look at it with me."

The next day, we carried bins and boxes to her large covered patio and spread parts of her collection out on a long table. The rainbow-colored rectangles of lighters came first. I estimated about 250. The toothbrushes, bottle caps, and fishing floats (organized by size and type) we didn't count. The plastic umbrella handles—it took Noni some convincing to haul all those out. They covered the entire ten-foot table with drab black, white, and brown hook shapes, a few faded greens and reds. We divided the task of counting and came up with 591. I asked Noni what she planned to do with them, but she just shrugged. She brought out a large tote full of plastic fragments that she had cleaned. I ran my hands and arms up to the elbow through the colorful rubble. It felt good, waxy, and smooth to the touch.

Noni is discerning about what she collects, otherwise she'd be buried in trash. She had two objects she wanted me to see in particular. She had found both items after the earthquake, tsunami, and nuclear meltdown in Fukushima, when some of the tons of debris washed away by the disaster started appearing in Hawai'i. She connected these items to that suffering. One was a blue, yellow, and white striped volleyball, dented and worn, with faded drawings in marker covering one side—hard to tell what they were, maybe animal faces. The other was a young girl's pink flip-flop, which had some bite marks out of it but was otherwise intact. It had Japanese writing and a picture of Sailor Moon and her magic cat, who endowed a regular schoolgirl with the power to save Earth from the forces of evil.

When it happened, the Fukushima disaster didn't deeply register with me. I slipped into numbness, evaded the news stories. When Noni showed me the weathered volleyball and pink shoe, my protective screen still held. I didn't feel the same charge that she did.

Later, I read a piece by Richard Lloyd Parry in *The Guardian.* He lived in Japan and covered the earthquake and tsunami as a foreign correspondent. He felt, even as an eyewitness and professional chronicler, that he had failed to grasp what happened. He quoted another journalist who wrote of the challenge of conceiving such events, and "the peculiar necessity of imagining what is, in fact, real."

I read this not as defining the bounds of some objective reality, but as coming to understand what a chain of events might have meant, and why it mattered—for all entangled by it, for all before, and for all after.

Parry focused on a single elementary school in a rural village called Kamaya, two hundred miles north of Tokyo. When the earthquake struck, the teachers and administrators at Okawa School efficiently followed protocol and evacuated within minutes. Parry describes schools in Japan as some of the best prepared for handling disasters, with procedures and drills for earthquakes, fires, and floods.

For a variety of reasons—some still inexplicable, but perhaps partly because they also could not imagine what was happening—the adults did not take the next evacuation step and gather the children to escape the tsunami by climbing a hill right next to the school. They ignored calls from city officials to flee, and overruled children who wanted to run to high ground. Townspeople who made it to safety on the hill recalled hearing the cries and screams of children as the wall of black water swept over them.

Later, some parents sued. They arrived for the verdict carrying a banner with the pictures of their children and the words, "We did what our teachers told us." In the end, writes Parry, "out of 18,000 people killed by the earthquake and tsunami, only seventy-five were children in the care of their teachers. All but one were at Okawa Elementary School."

This brought the sandal home to me. Like the albatross carcass, like so much else, I resisted bringing it close. But how do I live unless I can recognize—re-know—the shards of the present?

Sailor Moon gains increasing powers over the years of her existence in manga, movies, and TV series. She transforms into Super Sailor Moon to stop a villain from destroying Earth, and finally into Eternal Sailor Moon, capable of resurrecting with her love her fallen Sailor Soldiers.

My only power is this: to practice "hauntology." To embody and record our pained entanglements. To imagine this reality and remember—re-member—possible futures.

BACKYARD

Heidegger gave his lecture called "The Thing" before an audience of wealthy businesspeople in Bremen in 1949. It marked his first public appearance after the war: He was banned from teaching in the universities because he had been a Nazi.

Murder is the word Stan Ulam uses to mark his friends and colleagues lost—not killed, murdered. It becomes a refrain in his book: so-and-so murdered, and the date and place if he knows it. "Stozek...murdered by Germans in 1941." Łomnicki, the mathematician: "He too was murdered by the Germans in Lwów in 1941...Schnauder was murdered by the Germans in 1943... Ruziewicz was murdered by the Germans on July 4, 1941...Schreier was murdered by the Germans in Drohobycz in April 1943."

An accounting,

every one. Often parenthetical, the

reference to the murder, the parentheses that held

Ulam's whole life.

Of his family

he writes only, "I had no evidence

that any...survived." His father, uncle,

sister, her child. He writes of them

not a single word more. Too

total, perhaps, too

suffused through the body and all

its processes, this grief,

which means weight.

: :

In the summer of 1935, as Hitler's SS arrested Jews, activists, queers, and others they considered "threats," Heidegger delivered a lecture series to his students at Freiburg called "Introduction to Metaphysics." Darkness erupts into this work. Heidegger sees around him only death, decline. The gods have fled, technology has conquered the earth and reduced people to interchangeable units of value.

Heidegger wanted to go back to a "before," to a past in which humans had a more authentic relation to Being, before Plato. But he put some kinds of beings above others. He believed the German culture had superior qualities that could spark a metaphysical revolution.

Europe, he told his students, "lies today in the great pincers between Russia on one side and America on the other," both nations caught "in the frenzy of unchained technology." He repeats: "We lie in the pincers. Our people, standing in the center, suffers the most intense pressure—our people, the people richest in neighbors and hence the most endangered..."

For Heidegger, to be entangled, to be rich in neighbors, meant to be vulnerable, from Latin *vulnus*, wound, capable of being wounded. By 1949, not only contiguous neighbors—those actually nearby—created the danger. Technology—whose ultimate realization was the hydrogen bomb—had made distance disappear. Everyone was a neighbor.

The European refugees and U.S. citizens who built the bomb hoped this same technology would save them. Edward Teller had thought through the implications, in more scientific terms, years before Heidegger. The physicist Robert Serber remembered seeing on Teller's blackboard at Los Alamos during the war a list of ideas for weapons and their potential for destruction. "For the last one," Serber wrote, "the method of delivery was listed as 'Backyard.' Since that particular design would probably kill everyone on Earth, there was no use carrying it elsewhere."

SORRY

Yukiyo said she wanted to ask me about apologies. We had met to talk about grant applications, plans to collaborate, usual stuff. We had finished up the business and I needed to leave. I was in a hurry, as always, but Yukiyo said, "I want to ask you just one more thing." She talked about another poet she worked with on a project regarding the poisoning of the Willamette River, here in Portland, with PCBs and other industrial chemicals. Yukiyo suggested to this poet, a white woman like me, that she might apologize to the river, to the Chinook salmon, to native communities, to all harmed by white colonizers. The woman said she couldn't. She said there is not a word in our language for that kind of apology.

Yukiyo wanted to know what I thought about this. We spoke about apologies and what they mean, about how the word "sorry" is misused to manipulate others or evade actual accountability. We talked about what an authentic apology might be, and what it might accomplish. Yukiyo resisted the idea that apology might need to accomplish anything, or anything we could measure. "What if it's just something that happens inside, a way of being?" she asked.

We talked about existing as both victimizers and victims, and all the ways those states reverberate through our lives. I have benefited my whole life from money that supports nuclear weapons. It paid for every bite of food that entered my mouth as a child, every piece of clothing on my back, all my security and privilege. Yukiyo asked me, would I, if I were performing in Los Alamos, apologize—to all those displaced and harmed by the nuclear lab.

Something in me contracted against this; it felt impossible. "I'd have to apologize for...well, it would be a global apology, for everyone harmed all around the world by nuclear radiation, to all beings," I said.

Yukiyo then told me something I hadn't known about her partner, Yu Te. His grandparents came to Cambodia to escape both the Japanese Imperial Army and the Chinese Communist Party's brutality. During the Vietnam War, when Yu's mother was 12, the U.S. bombed her village in Cambodia. A bomb hit her school and she saw classmates die.

"Every time I am with Yu Te's mom, I feel sorry inside," said Yukiyo. "For all she suffered then and since. I'm in a state of being sorry."

I asked her if she'd ever considered speaking her apology. She said, "Yes. I almost feel ready. We have been building more and more trust."

And that's when it struck me. I'd been sitting here, with this friend I love right in front of me, speaking intellectually of global apologies for nuclear technology, and never once considered apologizing to her. Not in all the years we've known each other. I sat up straight in my chair and forgot to be rushed and impatient.

"Yukiyo," I said. "I'm sorry. Your family has suffered so much from nuclear weapons and war. You've lost your mother, your uncle, your aunt, and had such hurt and fear in your family, and in your own life." Yukiyo—she looked in pain. Tears came up in her eyes and spilled down her cheeks. I put a hand on her arm, tears came up in me also. I didn't know fully what to say. I fumbled—it's not as clear as I'm writing it here—but the words came out, from my body, my heart and my guts, not my head. I felt it all through me, that state. The state of being sorry. It hurt, because it required really seeing Yukiyo, and knowing and feeling how she has suffered. It required seeing myself also, or feeling myself, and the harm my privilege carries. It felt like release, like something broke free in me.

We sat looking at each other, a little overwhelmed and stunned. A little awkward. Someone interrupted and pulled us from the moment, which was probably a relief to us both. Later, Yukiyo wrote me this email:

> Your apology came by total surprise (maybe it surprised you too?) I didn't have time to prepare its coming! Thank you Allison for being so raw and so real. We are really present at that moment, even as the rest of the world was worrying about tomorrow or angry about yesterday, my feet were completely grounded and your eyes are my gravity. ❤❤❤❤

Yukiyo told me, in the moment, "thank you." She did not offer forgiveness, and I did not ask for it. It was a way-of-being kind of apology.

BUTCH

Lance Christman never met his father, the PBY pilot Elwyn Christman. I located Lance in the summer of 2011 after calling the dozen or so Christmans I found listed in or near Mt. Angel, Oregon. No one returned my calls.

I learned online that Elwyn had married a woman named Celia Gow, whose last address was Port Hueneme, California. She died on June 6, 2011, weeks before I found her. The phone number at that address connected me to Lance's ex-wife, who led me to Lance. He sent me two CDs full of letters and photos from his father.

I also emailed a veteran who posted in 2002 that he had written a history about the last squadron Christman commanded. Raymond Hurlbut answered my email almost instantaneously. He had served as Christman's radioman, in charge of communicating with base and keeping their bearings at sea, when the squadron was stationed at Roi-Namur in the Marshall Islands during 1944 and 1945.

Hurlbut was the only person I spoke with who knew Christman while he was alive. I interviewed him via Skype in August 2011, when he was eighty-nine years old. He was animated and full of energy. Golf clubs leaned against the wall behind him. At one point, he leapt out of his chair with surprising alacrity to show me his World War II veteran baseball cap. "See it? See it?" he said, jamming it on his head and grinning.

"I remember when [Christman] got the news that he had had a son," Hurlbut said. "He was very excited about it."

During the early part of his naval career, Christman proclaimed his proud bachelorhood in letters home. In 1939, he wrote from the Philippines to "the Hound Dogs" (probably his cousins), declaring, "I'm going to be the ideal batchelor [sic]." In a 1940 letter to his sister, he reported meeting “the cutest little blonde” he planned to take to *Gone With the Wind*, helping ease his pain from the departure of “the red head” gone back to Hawai‘i.

After Christman survived his plane being shot down at Jolo and those two weeks making his way back to base after the Navy declared him dead, the military sent him back to the states. He spent more than a year as a flight instructor in Jacksonville, Florida. He wrote home about meeting a “Red-head” on a trip to Dallas and telling “June” not to “sit around holding her hands” waiting for him.

By October 1943, he was in San Diego. On February 24, 1944, he described a furlough in the city with “plenty of liquor and women,” but something else happened: He met Celia. She wrote the next letter to the cousins, dated March 15, 1944: “Well, as you no doubt know by now, we are married...I was so nervous I didn’t know whether I could stand it or not, and Cris [sic] was the same.”

They found an apartment at the Naval Air Station in Crows Landing in the San Joaquin Valley and lived together for about three months. Celia, or “Cissy” as she signed herself, continued the letter writing. In April 1944, she wrote the cousins: “Haven’t been doing much except keeping house and cooking. I hate to cook. And I don’t understand how 2 people can get a place so cluttered up.” Later she wrote that there were not enough airplanes for training. “Cris plays pool all day at the base.”

In early June, the Navy promoted Christman to Lieutenant Commander in charge of his own squadron, the VPB-133. “The first thing he did, he had a party,” Hurlbut told me. “It was for all hands, enlisted men and officers to get acquainted.” Hurlbut said it was a joke among the men that Christman picked his plane crew that night based on who could hold their liquor. “That was quite a wild party,” Hurlbut said. “That was the first time we met him and the next day we found out we were on his crew.” Hurlbut was twenty-three years old.

In late June 1944, Christman's squadron shipped out to Hawai'i. In July, he wrote to his cousins: "The Christmans are expecting a visit from the stork." He reported buying pounds of coffee and bottles of bourbon for the officers and crew to keep them stocked wherever the military sent them.

In September, the squadron headed to the Marshall Islands. Their assignment: to bomb Japanese ships, as well as Japanese-held bases on some of the Pacific islands that the U.S. military had bypassed in its push toward Japan, a "mop up" mission.

"You mentioned that Christman saved your life," I said to Hurlbut. "Tell me about that."

"We were doing a bombing raid," he said, "and one of our engines quit. We didn't know what happened, but we lost an engine. Through his skill at flying and the cooperation of the rest of the crew, we flew that plane back, I think it was about two hundred and sixty miles.

"At times we were not much more than fifty feet off the water. I'm communicating with the base and telling them our progress, and when we came in for a landing, the whole, excuse me [his voice breaks], the whole runway was lined with guys cheering, because they all heard we were coming in on one engine and you know not that many people thought that plane could fly on one engine, but he did it."

: :

Christman's letters to Celia from this time are full of tenderness. He calls her "honey" and "dear" and asks constantly about her health. "Darling you are just the sweetest wife a man ever had," he wrote her in December 1944.

He writes of almost nothing but the coming baby. "By the way, has Butch been kicking you around much lately?" he asked in September, and wondered what they should name him. In

October he wrote, "Sweet, just when does the doctor think this little Butch will be putting in an appearance?"

I only have Christman's letters after he deployed, not Celia's: His letters to her are a dialogue with silence. It seems she suggested some names for the baby. He liked Vance but rejected Candace because he thought it was too long and complex. "Where in the world did you get that name from?" he asked.

He was fixated on a boy. "Hon, you better get him a pair of boxing gloves so I can start right in teaching him the manly arts," he wrote in November.

In December, he reported, "Have about $90.00 bet on Butch's sex so see that it's a boy! We really can't afford a girl. They cost more than boys."

At the end of December, he chides Celia: "Now see here my sweet. This is no time to be talking about having cold feet. You hear me now? It might even give Butch an inferiority complex!"

Lance was born February 1, 1945. Christman's last letter is dated March 20. It is typed, which is unusual—most of his letters were handwritten. He still refers to Lance as "Butch," and repeats that he plans to get him boxing gloves and teach him "the manly arts" as soon as he gets back. Christman writes that he has heard his relief is coming, and that he should be home by the middle of April. The next letter in Celia's collection is a note from the Vice Admiral of the Navy expressing condolences on Christman's death.

: :

Christman, his crew, and five other airplanes had been sent to Iwo Jima on March 23, 1945. Their job was to patrol the ocean north of the island and attack small Japanese boats called "picket" boats that were keeping watch for B-29s headed to bomb Japan. On March 30, Christman and his crew went out on patrol.

"We spotted this boat, and we made a run at it," said Hurlbut. "We could see them firing, small fire, at us and we just ran out of ammunition and dropped our bombs. I don't think we damaged it enough that it would sink, but Christman thought it might be disabled. When we flew back to Iwo Jima he wanted to load up some more ammo and get gas and get going to attack again.

"You have to understand Iwo Jima was an Army airfield, and we were just visitors there. The Army ground crew told us, 'No, you can't take off until we check the plane because you've got holes in it. We don't know what kind of damage it is.' Christman said, 'Well, I flew it back here, I ought to be able to fly it off again. We've got a ship that I can sink.' The guy said, 'You have to get permission from the duty officer,' who was in a little tent along the runway.

"So Christman went down there to get permission. An Army fighter plane that had been out on a raid came in with both wheels locked and swerved off the runway, headed right to where these guys were standing. So they ran for cover and he tripped and fell down and the plane hit a truck and knocked it right over on him."

A few years after Christman died, Celia moved with Lance to Christman's family farm in Oregon to take care of his eighty-year-old uncle. Lance grew up on the same land as his father. "Living on a farm in Oregon is the only way to grow up as far as I'm concerned," Lance told me. He remembers digging tunnels through hay bales and shooting at targets with a .22 rifle. "One day I shot a robin. I went over and it was still kind of flopping around and I thought, 'That's it, I'm not killing anything else.'"

Lance spent much of his career working for military contractors building projects for the Navy in the Pacific. He had seen the albatross on Midway. I asked him what he was working on, but he said he couldn't tell me. He also owns an airplane, a Piper Cherokee, and has had a pilot's license since he was about twenty. "Is that because of your dad?" I asked. "Not necessarily," he said.

"I just had the bug. And probably because my mom talked so much about flying. She didn't get her license, but she was learning to fly, and she did some acrobatic training."

Lance believed his mother had never remarried. But when she became ill, he went through her papers and found a marriage annulment from not long after his father died; she had married a second time for only a few months. She worked at the Social Security Administration her entire life, bought a house, and raised Lance on her own.

Christman is buried in the Fourth Marine Cemetery on Iwo Jima. Lance sent me a photo provided by the military: rows and rows of identical white crosses stretching off into the distance.

DECLINE

I asked Yukiyo to try and contact Kunimichi Odagaki, designer of the Honda Odyssey. She managed to reach him after many phone calls and emails. Retired, he seemed to be teaching at a school called Odagaki-juku, which Yukiyo says means "Odagaki-cram." Its curriculum focuses on how to succeed in the corporate world with attributes like being "tough but creative and flexible."

Yukiyo was surprised that Odagaki seemed cooperative and happy to talk. He asked for questions or the content of the interview in advance. I wrote a brief, sincere letter, outlining my journey with the car part, my discovery of him and the story of the Odyssey, my trips to Hawai'i, and my encounters with plastic. I included a few questions about his thoughts regarding the issue of plastic, and whether he had any advice for me about what to do with the car part. "I can't recycle it here," I wrote, "but do you think that if I brought it to Japan I could recycle it there?"

Yukiyo translated my letter into Japanese and sent it off to Odagaki. He replied the very same day, on September 11, 2018. Yukiyo translated his response: "Reading the letter that I sent to him, he realized that the questions are not about science or technology (which he will be more comfortable talking about). He doesn't think he can answer these questions using his limited skills. He wrote, 'I regret that I need to decline the offer at this time.'"

MIKE

The thought that came to Stan Ulam as he stared out the window that winter day in 1951 remained secret for a long time. The U.S. government still considers its details classified.

The first person to reveal Ulam's breakthrough was Howard Morland. He joined the Air Force in 1965 because he wanted to be a pilot; he expected the war in Vietnam to be over by the time he finished training. Instead, he found himself flying transport planes across the ocean, bringing supplies to troops and returning, as he writes, "with dead bodies in aluminum boxes."

One day Morland received training on transporting what the military called "special weapons"—nuclear bombs. Their small size shocked him (aluminum containers eighteen inches around and six feet long), smaller than an average coffin.

He became increasingly obsessed with nuclear weapons. He felt startled by most people's indifference to the fact that the U.S. and Russia had thousands of bombs that could wipe out entire cities. He thought people didn't understand the threat they lived under, because a hydrogen bomb remained something exotic to them, beyond their ability to imagine.

Morland decided to make the thermonuclear bomb real for people. He wrote: "I wanted to prove with words and line drawings that nuclear weapons have size and shape. They occupy space. The parts are made by real people using real industrial equipment." He set out to uncover the inner workings of the H-bomb.

: :

At Los Alamos, by 1951, Ulam had turned away from work on the Super. Nearly a decade after he arrived, the scientists were no closer to creating this weapon, and Teller and Ulam had grown to deeply dislike one another. Teller fixated on a design that Ulam had proved would not work. Teller's version packed a fission bomb with layers of fuel in an attempt to get it hot enough to spark a thermonuclear fire, but the math carried out by human and machine computers indicated it would blow apart before it reached a high enough temperature.

Sitting in his living room that afternoon, Ulam suddenly saw how to overcome this problem. Why not separate the bomb into two stages? Place an atomic bomb at one end of a cylinder, and the huge flux of radiation coming off the bomb would heat and compress the thermonuclear fuel at the other end, forcing its nuclei to melt together, creating fusion.

Ulam brought his idea to Teller the next day. Despite their animosity, Teller instantly saw how the new design could work, and improved on it. The two wrote a still-classified paper on a two-stage Super. Now known as the Teller-Ulam configuration, it forms the basis for nearly every thermonuclear weapon on the planet.

None of this was public knowledge in 1978, when Morland received a contract from the magazine *The Progressive* to publish a series of articles on nuclear weapons. But Morland was relentless in his quest. He pored over the scientific literature and crisscrossed the country, visiting weapons plants and talking to scientists. He developed the trick of speaking with authority on classified details he had only guessed at, lulling experts into thinking that the information must be public.

The Progressive published Morland's article, "The H-bomb Secret," in April 1979. The U.S. Government sued the magazine and confiscated all the copies. Six months of court proceedings enabled Morland to keep talking to scientists and gathering details. He sensed he had gotten the basics right, but something about the design he proposed bothered him. He knew the weapon had two stages, and he knew the radiation from the atomic bomb compressed the thermonuclear fuel, but he wasn't sure exactly how this mechanism worked.

Eventually, the government censors themselves revealed the answer. Morland's lawyers had to prove that everything Morland wrote existed in the public literature. To do this, they gathered testimony from scientists regarding the accuracy of Morland's article and the sources he had relied on. The scientists duly pointed out Morland's errors. The government censored the briefs before their public release, but they left one crucial paragraph intact, confirming that Morland had indeed gotten the role of radiation in compressing the fusion fuel wrong. His design was missing a crucial element: plastic.

Radiation from the plutonium bomb could heat the thermonuclear fuel, but the original designers worried the implosion would blow the assembly apart before the fuel could get hot enough. They needed some other material to intervene, a simple material, of low atomic weight, that would not interfere with the reactions. They chose polyethylene—just carbon and hydrogen, made from the bodies of ancient creatures, out of the same molecules that make up every living being.

The Los Alamos builders nailed thick slabs of polyethylene inside a cylinder-shaped casing that contained deuterium fuel. By 1952, polyethylene had entered consumer culture in the form of Tupperware and squeeze bottles. Soon it would become ubiquitous, the most common plastic on the planet.

: :

The same year the scientists lined their H-bomb with polyethylene, DuPont and Imperial Chemical Industries lost their their ability to control polyethylene production and prices in an antitrust judgement in the U.S. courts. Other U.S. companies soon built plants, glutting the market, causing the price to tank and driving the need for new products to use up all those polymers.

In 1956, the hula hoop swept the nation, one of the first fads of the baby boom generation. By 1958, polyethylene hoops sucked up one-third of the nation's production of a new high-density polyethylene, raw plastic that otherwise might have gone unsold. The industry targeted other products to children: polyethylene toy soldiers in Army green, and, for girls, candy-colored "pop" beads with pegs and sockets for necklace-making.

Today, U.S. manufacturers produce enough polyethylene every year to equal the weight of every person living in the United States, about 44 billion pounds. The various types of polyethylene appear in everything imaginable: mostly packaging—including water bottles, plastic bags, shrink wrap, and blister packs—but also buckets, garbage bins, carpeting, clothing, water pipes, replacement hips, bulletproof vests, and the car part that ended up one day tangled against my fence.

: :

The polyethylene in the original hydrogen bomb worked exactly as intended. When the atomic bomb went off, it heated the plastic to a million degrees in an instant, creating a plasma that expanded explosively, squeezing the deuterium and sparking a thermonuclear fire.

The result of Ulam's idea that day was the world's first thermonuclear device, code-named Mike, perhaps because the "M" referred to the expected yield in megatons, or millions of tons of conventional explosives—the largest explosion ever created by humans to that point.

Ignited at Enewetak Atoll in the Marshall Islands on November 1, 1952, Mike exploded with the force of nearly a thousand Hiroshima bombs. It vaporized the island of Elugelab, leaving a dark crater in the tropical waters and lifting tons of coral and sand, the remains of sea creatures, into the atmosphere. Alive with deadly radiation, the tiny particles drifted around the world and fell back down.

Every being on Earth now carries radiation in its tissues from the nuclear testing of the 1950s. Scientists have figured out how to use this radioactive trace to determine how long something has been alive—a tree, or a whale, or an individual human cell—because those born later have lower concentrations as the radiation dissipates. This mark in the cells is called the "bomb pulse."

THE LIVING

After Jen and I split, I lived nomadically for a while: in the places of different friends, in an apartment I rented for a year. The car part came with me, one constant. It stayed in the back of my car or made its way into new corners. During the years of writing, I also traveled and gave readings and talks about plastic, at conferences and colleges, in Washington, D.C., California, New York, Colorado, Philadelphia. I flew, adding to the carbon warming the atmosphere, and sometimes I brought the car part, folding it to fit into a bright red roller bag. I brought it out and propped it on podiums and lecterns as I spoke, and it made its flat "clap" sound, nonresonant.

My favorite parts of these trips were the workshops with students, in which we collected plastic and wrote and talked about it. Each time it amazed me that a group of strangers would consent to pick up garbage with me. Some refused. One participant at the University of Hawai'i sat out the plastic collecting; she probably knew we'd find swarms of cockroaches. At Washington State University, we dumped our collected plastic on the floor and dozens of ladybugs emerged and circled the room as we talked, alighting on furniture, belongings, people. Some of the students seemed amused, some recoiled.

Mostly people took on this interaction with garbage as a legitimate activity, even though my own reasons for collecting plastic remained hazy. I came as an invited guest, imbued with authority, which originally meant "father, one who creates, the person responsible." The plastic itself was numbingly familiar—the same detritus of U.S. consumer culture everywhere—but collecting with other people felt more meaningful, less lonely. We could marvel, gross out, mourn together. It was the living, not the plastic, that drew me.

III.
THE LIVES

FREEPORT

I arrived in Freeport in the middle of a freak ice storm on the Texas Gulf Coast. I traveled south on 288 from Houston, nearly alone on the road in my rental car. Outside the city, trees and farm fields lined the highway, a lush world of green, empty of people, caught in ice. I topped a small rise. Before me spread a smoking landscape of factories: grey stacks lit here and there by spurts of fire.

Nearly two dozen plants surround this town of 12,000 people. The behemoth of these: Dow Chemical, the world's leading supplier of polyethylene. The Dow complex in Freeport—this vista of cement, metal, steam, and flame—makes up "the world's largest integrated chemical manufacturing site," as the local chamber of commerce proudly proclaims. It sprawls across an area nearly the size of Manhattan.

Dow came to Freeport in 1940. It built a factory to suck water from the Gulf of Mexico and used its patented process to extract magnesium, a kind of salt.

Magnesium burns white hot and refuses to go out. Water doesn't quench it; magnesium just pulls the oxygen out and burns hotter. The U.S. used it in firebombs, like the ones that Curtis LeMay and his pilots dropped on Japanese cities in World War II, ending hundreds of thousands of lives.

I met Melanie Oldham in Freeport. I had contacted her by email after finding her name online. She is one of the founders of Concerned Citizens of Freeport and has been fighting pollution in the area for years. She picked me up at the Holiday Inn Express across the street from the Dow and BASF plants. "We call this Chemical Alley," she said when I climbed in the car, waving to the row of budget hotels. Because of prevailing winds, pollution tends to settle here, she said. I'd been noticing an acrid, sweet smell since I arrived.

I'd expected Melanie to drive me around the factories, but instead she headed north ten miles to Lake Jackson, the bedroom community Dow built for its employees. We drove through wandering tree-lined streets with names like "This Way" and "That Way," past neo-colonial mansions fronted by big lawns. The town of 24,000 had a feel of sleepy prosperity, a skin of timelessness.

Lake Jackson takes its name from Abner Jackson, who owned one of the state's most profitable sugar plantations during the nineteenth century, powered by the labor of two hundred enslaved people, making him one of the state's largest owners of human beings. The Lake Jackson Historical Association gives tours of the plantation site once a month. The Brazos Mall sits near the site where enslaved Black people lived, and where some are buried.

Dow and other plant executives live in Lake Jackson; the estimated median household income in 2016 was $78,000. Plant workers live in Freeport, closer to the plants and to pollution, with a median income of $37,000. Lake Jackson is majority white. Freeport is 65% Latinx and 11% Black.

The factories that surround Freeport send a mix of gases and particles into the air. Of the ninety plastic plants in the Greater Houston area, Dow is the third top emitter of nitrous oxide and pollutants called volatile organic compounds (VOCs) that cause smog. The most common VOC in Freeport is ethylene, the raw ingredient for polyethylene. Dow accounts for the vast majority of this pollutant.

Smog consists mostly of ozone—three oxygen atoms strung together. Ozone is called toxic, which means "imbued with poison" and comes, perhaps, from the root for "to flee." When a person breathes in ozone, it breaks apart molecules in the lungs. The body responds to this injury by sending white blood cells; the tissue turns red and swells, like a sunburn. It is "inflamed," as in on fire.

Asthma refers to the constant state of this burning. The lungs are always on fire. When pollution enters, the lungs inflame even further. They swell, making it hard to get air. Pollution is called a "trigger" for attacks of asthma. So is dust, cockroaches, the weather, including thunderstorms, and strong emotions like anger, fear, excitement, laughing, yelling, or crying.

Ozone does not only do its damage in the lungs. New science shows it also travels into the blood where it can cause the whole body to become inflamed. It can knock the heart off its rhythm, narrow arteries, raise blood pressure, and block blood flow to the heart and brain. This snuffs out cells and sets off heart attacks and strokes, which kill more people in the world than anything else, including infections and car accidents.

Ozone can cause these harms at levels below those allowed in the U.S., though much of the country violates these limits anyway. Brazoria County, where Freeport sits, has some of the highest ozone levels in Texas. It has never met the federal limits for ozone set in 2008, much less the lower levels proposed in 2015 on the advice of scientists, which the state's own environmental agency fought against.

Melanie got involved in pollution issues because her twins both had asthma, and her daughter's was severe. "People would talk about pollution in Brazoria County, and I couldn't believe what I was hearing," she told me. "Then I started to do my own research."

Melanie is tall and blonde, with light, intense blue eyes. She works as a physical therapist, traveling to patients' homes, many of whom worked at the plants and are now sick, she said.

Industrial plants receive "permits" from the state environmental agency, the Texas Commission on Environmental Quality (TCEQ), to send pollutants into the air and water. The state's lax permitting process gives the plants lots of leeway to pollute, and relaxed enforcement and low fines mean that companies often exceed their pollution limits. “My ex-husband worked for Dow,” said Melanie. “I overheard the conversations. The companies planned to be fined. Instead of fixing the problems, they just worked the fines—pocket change to them—into their budgets.”

According to the Environmental Integrity Project (EIP), nearly two-thirds of the plastics plants in the Houston area violated air pollution laws from 2015 to 2018. A loophole in the law allows them to pollute even more: When plants experience what they call “upset” events like malfunctions and the need for unplanned maintenance, they claim the right to release more pollution than usual—without penalty. Of all the plastic plants in the region, EIP reports that Dow released the highest amount of pollution from unpermitted “upset” events during that time period: more than three million pounds.

Melanie and I ended up that January day at On the River, a restaurant where we ate shrimp and catfish from the Gulf. I noticed that Melanie’s black-and-white plaid suit jacket had a rhinestone cross on the back, with angel wings above it. “I hope you’ll write about us in your book,” she said. “We could sure use some help.”

By the time Melanie dropped me back at the Holiday Inn Express in Chemical Alley, it was dark. I felt restless. I still didn’t have a sense of the plants. Even here, in the heart of the petrochemical industry, it all seemed inaccessible, abstract. I’d never seen anything like the sprawling landscape of factories; I was having a hard time imagining this reality.

I got into my rental car and started driving around again. The Nolan Ryan Expressway was designed to speed the commute from workers’ homes in Lake Jackson directly to the plants, so there were few exits. The old road went right past the plants, but it had no shoulder—the only side roads led into the plants themselves, toward floodlights and security gates.

There seemed no safe place to stop and take in the view, so I found myself looping around the same highway cloverleaf over and over, sailing slowly past the facilities, like massive ships alight in the dark, steam and flame billowing out of their stacks.

LIGHTS

On my second trip to Freeport, Melanie brought me to meet Jessie Parker in Freeport's East End neighborhood. We gathered on a sweltering July day around Jessie's glass dining room table. Jessie is a nurse who moved back into the home she grew up in a few years ago, after more than two decades away. Manning Rollerson, another East End resident, joined us. Jessie and Manning are activists and leaders in the historically Black East End neighborhood, and part of Concerned Citizens of Freeport with Melanie.

Jessie set out a tray of veggie snacks, left over from a family Fourth of July party the day before. She handed us each a cold water bottle. Her house was cool and somewhat dark, with white curtains closed against the heat. A TV played without sound beside the table. Cobalt and Aster sat on a couch against the wall, facing the silent screen. They had joined me for the trip: two weeks of driving along the Gulf Coast, then up to Lincoln, Alabama, to bring my car part to the Honda Odyssey plant.

I'd been thinking about this trip for years, and finally made it happen in the summer of 2019. I packed the car part into its red roller bag and we flew to my father's house in New Mexico to cut some miles of driving off the trip. In the end, we traveled more than 3,500 miles from Albuquerque to Alabama and back. Including the flights, we sent around 5,000 pounds of planet-warming gases into the atmosphere.

: :

I felt awkward and stiff at first, sitting at Jessie's table. I worried about Aster getting bored and acting out. I worried about what this group might think of my queer family. I can pass in the world as straight. When I was married to Jen, we often didn't register with strangers as a couple—people would hand us separate checks at meals. Once, boarding a plane, Jen made some joke to the flight attendant that I don't now remember. The attendant said, "You better not tell your husband that!" I stood right behind Jen. It flashed into my mind to say, "*I'm* her husband." But I didn't. I have been trained to not make a scene. To keep the peace.

Cobalt often doesn't have this choice. Their gender does not fit easily into any binary. Strangers sometimes stop them from going into what they believe is the "wrong" bathroom. It did not occur to me before this trip that in traveling with my new family, meeting with communities, I'd be constantly coming out.

Sitting at Jessie's table, I started out by accounting for myself, explaining my purpose in coming to Freeport: that I was writing a book about plastic, that I wanted to connect the widespread concern about plastic waste with the impacts on communities where plastic is produced.

"I'm interested in what you have to say," Jessie said. "We definitely could use help and support, so if this will help us..." she raised her hands, palms up, with a shrug.

Miss Jessie's shrug—I came to call her Miss Jessie—carried a history. I was not the first stranger to take an interest in Freeport. Activists in the community have engaged over the years with journalists, researchers, environmentalists, and documentary filmmakers. Also with lawyers purporting to help.

Freeport's East End is a product of segregation, established in 1930 by the town's white leaders as a "Negro reservation." The town forced any Black person not living as a "bona fide servant" at the back of a white person's property to live in the East End.

The community is bordered to the west by the railroad tracks, to the south and east by Port Freeport and its industries, and by the ship channel to the north.

Across the channel sprawls part of the Dow facility. In 2017, the company fired up a new plastic plant—the "crown jewel" of its

$6 billion expansion on the Gulf Coast—making it the largest ethylene plant on the planet, according to the company.

The Dow plant is called a "cracker." It takes molecules of ethane from natural gas and heats them to high temperatures, "cracking" them apart to form ethylene for plastic. It was the first in a $150 billion boom of chemical plants along the Gulf Coast, as companies looked to profit from transforming the oil and gas freed by fracking into products for the global economy. In the Houston region, the industry had forty-eight new plants or expansion projects planned as of 2019, adding to the 55,000 tons of pollution already released into the air every year by plastic facilities.

Industry giants, including those associated more with oil, like ExxonMobil and Shell, put big bets on plastic as their future profit maker. Projections show demand for gas and diesel dropping as vehicles get more efficient and electric cars take over the market. Analysts expect demand for plastic, on the other hand, to surge, thanks to growing economies in Asia producing a new class of "consumers." This increased purchasing power would drive demand for food, beverages, and other items in disposable plastic packaging. People would need these conveniences because, with increased work and wages, would come, according to one industry analysis, "a lack of time."

Lack. This word did not exist in Old English. Its origin is uncertain, but it may come from a root that meant "to dribble, trickle." The privileged in industrial societies long ago accepted this tradeoff—money and objects instead of time to be alive, not working—content with the dribbles and trickles. *You*

will lack

for nothing.

: :

The rush for plastic plunder happened before a global pandemic swept the globe in 2019 and 2020. When the pandemic hit, demand for plastic tanked. Dow even shut down its Freeport plant for a time. But the industry pivoted quickly to shore up its main cash cow: single-use plastic. A group of companies including Dow asked the U.S. Congress for a $1 billion bailout.

In March 2020, the largest industry association sent a letter to the Trump administration calling for a reversal of plastic bans and asking the federal government to declare single-use plastic bags "the most sanitary choice," despite a study in the *New England Journal of Medicine* finding that the COVID-19 virus lives longer on plastic than cardboard or cloth. The PR worked. California, Oregon, and other states reversed their bans on plastic bags.

The head of the Plastics Industry Association issued a statement declaring, "Single-use plastics can literally be the difference between life and death."

But whose life and whose death?

"She says she wants to know about the environment," said Manning Rollerson that day at Miss Jessie's. He looked at Melanie, the other white woman at the table, instead of at me. He wore a fatigue jacket with a patch that said Army on it; I would learn later that he is a veteran of the National Guard. He spoke so quietly at first that my recorder mic barely picked him up, but as our conversation continued, his voice grew more resonant and emphatic.

"We live this over here. Every morning you wake up, there's always a powder on your car. That's been going on for decades. The reason why people here have asthma and all kinds of chronic diseases, it's because it's here, it's right across the street. It dumps in our water, it's on our doors, it's on our cars. And each time what do we get? 'Oh, it's not toxic.' But we still have people sick."

"One Earth Day, I got Air Alliance, a group from Houston, to come down here," said Melanie. "They talked with the elementary students about ozone. I sat in on the classes. At one point they asked the kids, 'How many people have asthma?' I swear three-quarters of them raised their hands. The teachers said: 'They all have inhalers here, they keep them at the nurse's station.' And the thing is, they don't warn people, like they do in Houston. You don't know if it's a red alert day, to keep your child indoors. Or if you're elderly to stay in."

"Elderly like myself!" interjected Miss Jessie. She seemed far from elderly. She wore white capri pants, a black patterned blouse, and tortoise shell glasses. She calls herself "retired" but still works three days a week as a traveling nurse, after fifty years working in nursing in the Houston area.

"I'm graced as a person with two hands around the throat of life," Miss Jessie told me that morning, and as I came to know her in the months afterward, I understood what she meant. She is the person at the center of every parade and party, every struggle. She communicates directly with God and has access to a spiritual world as concrete as others' daily reality. She is a guiding force for her daughters, her four grandchildren, and others in her community, people she calls "spiritual daughters." She speaks with passion and ferocity about her patients. How she wept when she was head of nursing and one of her patients got a bedsore—"That does not happen on my watch!" she told me.

Miss Jessie got a message from God to move back home and help be the voice of her community. In the ten years since she returned, she developed asthma and relies daily on an inhaler, her lungs, like those of Melanie's daughter and the kids in the local school, swelling and burning in their effort to breathe.

Freeport is not unique. Across the country, a majority, 57 percent, of people of color live in counties with at least one failing grade for smog or particle pollution, compared with 38 percent of whites. Privilege keeps the impacts largely hidden from the white majority.

"For too long, one of the most invisible aspects of the plastics crisis has been...communities who live in the shadows and along the fence line of plastics refining and manufacturing," Carroll Muffett, the head of the Center for International Environmental Law, told *The Nation.* "These people are experiencing the impacts of our plastic planet in a way that is more immediate and more severe than just about anybody else in the world."

: :

People once called lungs "lights"—as in, the light organ. Lungs are mostly air. Delicate, hollow branches permeate their tissue, carrying oxygen to the blood, keeping cells alive. The obsolete meaning persists in the phrase, "I'll punch your lights out."

I'll punch your lights out.

In June 2020, people around the world filled the streets holding signs that said, "I can't breathe," the words that George Floyd—and Eric Garner and at least seventy others before him—uttered as a Minneapolis policeman knelt on his neck.

What can one say after such a sentence?

Perhaps this: Police violence is only one eruption of four hundred years of systemic racism and white supremacy taking Black people's breath. Floyd's murder happened in the midst of a global pandemic, which comes from the roots for "all" and "people," affecting everyone. But this virus, which attacks the lungs, does not have equal effects. Those most at risk are already sick. They suffer from diseases of the lung and heart, the kidneys, and liver. They endure diabetes and obesity; they lack healthcare, safe places to live and work, access to clean air, good food, and enough money to pay for it. In the U.S., those who are Black, Indigenous, and people of color suffer most and die more. In this sense, the virus did not make anything different.

These facts did not bring me to Freeport. When I visited, the virus and the uprising had not yet occurred. But the facts came with me anyway, out of generations of lives and the lights they carried. They came with me, and met me when I arrived; they demanded re-membrance, a future.

RISK FACTOR

Smog-related illnesses are not the only health problem for people in Freeport. As we wrapped up our conversation at Miss Jessie's that day, Melanie pushed across the table a study by the Texas Department of Health from May 2018 titled, "Assessment of the Occurrence of Cancer, Freeport, Texas, 2000–2015."

She had just succeeded in getting the state to release this to her after urging them to conduct the assessment, a follow-up to a 2012 study. The report is short, just seven pages. Using the Texas Cancer Registry, it compares rates of cancer in Freeport with those in the rest of the state. It concludes: "The number of all-age liver and intrahepatic bile duct, lung, nasopharynx/nose/nasal cavity and middle ear, and stomach cancers was above the range expected."

The study authors note that they performed the first two steps in investigating a cancer cluster as outlined by the U.S. Centers for Disease Control and Prevention. Those are: consulting the community and counting cancer numbers. The second two steps they did not carry out: conducting an epidemiological study and identifying potential causes of the cancer—or, as they call them, "risk factors."

Lung cancer is common. It is one of the top ten global killers. It mostly kills people like my mother who smoked, although air pollution and workplace chemicals also cause this cancer. The other cancers are rare in the United States, and they are all linked to toxic chemicals. The most common of them in the U.S. is liver cancer. The body takes chemicals in through nose and mouth, down to the lungs and into the blood where the liver tries to process them out through the "intrahepatic bile ducts," which just means small openings inside the liver. Since 1980, the incidence of liver cancer has tripled, and it is the fastest-growing cause of cancer death.

"Here's the bottom line," said Melanie. "We might have jobs at the plants, but then we get ready to retire and we come down with cancer. We are a sacrifice community. They've decided they are going to sacrifice some of us, and maybe our lives."

: :

In her book *Don't Let Me Be Lonely*, the poet Claudia Rankine writes that her editor wants to know why she is interested in the liver. The book has a graphic, a simple line drawing, of the liver inside a person, connected to the mouth by the tube of the esophagus, with the outline of the United States in the position of the intestines.

She writes: "The word *live* hides within it."

And: "It is the largest single internal organ next to the soul."

And: "Any kind of knowledge can be a prescription against despair."

Over the phone before we met, Miss Jessie gave me directions to her home on East Seventh St. "It's the only brick house on the street," she told me. It's also one of the only houses left. I could see it from blocks away as we approached through the mostly vacant lots. The house has green trim and a white cross out front that says, "He is risen!" It sits about five hundred feet from the port. The water isn't visible from the neighborhood, just ship cranes and low buildings, and a storage terminal that says in faded letters: "American Rice."

Between 1930 and 1970, the East End's Black community flourished, sustained by the shrimping industry and jobs at the nearby city works, the Port, and the chemical plants. Over the years, Latinx people moved in, and some whites. Manning and Miss Jessie described the neighborhood of their memory, interjecting and talking over one another. "Remember Miss Tiny? I used to stay with her," said Manning, pointing. "Mr. Jacques used to live over there, Mr. Goods over there."

"We had at least two churches on every street," said Miss Jessie. They list the churches: St. Emmanuel, Wesley Methodist, New Jerusalem, Second Baptist, the Holiness Church. Businesses, too. "You had the Three Brothers Grocery, the Tropical Club, we had barbershops, we had the lady who did hair on Fifth Street," said Manning.

"The community raised us," added Miss Jessie. "Everybody was your parent, and you didn't get away with anything. If one of the women saw me down the street, it didn't matter what I was doing, she would grab me by the collar and tell me, 'It's time to go home Jessie Mae.' We were well cared for. It was a great place to grow up."

For years, the Port has been buying up vacant properties in the East End and, my hosts told me, pressuring homeowners to sell for low rates, then threatening to take the property if they don't sell out. That's why Miss Jessie came back to the house her grandparents built, and where her grandmother raised her.

Port Freeport has big ambitions. Though not a huge player in foreign trade—it ranks 27th among U.S. ports—it aims to establish itself as the main entry for goods flowing to and from the U.S. heartland, according to the *Houston Chronicle*.

The word heartland did not refer to the middle of the U.S. until after World War II, when the region gained status as the geographic and cultural core of a mythic white, rural, farming nation. It served for whites as a "psychic fallout shelter" from nuclear-armed adversaries on the outside and growing, racially diverse cities on the inside, according to the scholar Kristin L. Hoganson. The myth of the heartland aims to erase completely the Indigenous societies that settler colonists pushed out and attempted to exterminate.

The first recorded use of the word heartland, in the seventeenth century, meant "a place where love resides." The early 20th century brought the geopolitical definition, originally applied to Eurasia, as "a strategic center of industry, natural resources, and power."

It is this vein Port Freeport wants to tap: The middle of the country still produces more exports than any other U.S. region. The Port plans to dredge its harbor channel from 46 to 56 feet, making it the deepest harbor in Texas. Port commissioners hope this will attract the super-Panamax cargo ships now able to pass through the expanded Panama Canal.

The Port's plan for the land where the East End now sits: warehouse buildings for new cars and other cargo. This reflects the hierarchy of values in global capital: stuff comes first; consumers—the people who buy the stuff and keep profit flowing—come second. The lives of those who might interrupt this flow have negative value. They are obstacles for removal.

Manning, Miss Jessie, and others in the East End refuse to disappear. They join generations of Black activists and leaders—many of them women—working for justice and better environmental conditions in their communities. Low-wealth neighborhoods and communities of color bear the brunt of the nation's hazardous pollution and toxic dumping, but the dominant white environmental movement has largely ignored and erased these activists, despite their success at connecting issues of environmental and social justice and mobilizing entire communities. Their life-and-death struggles remain outside the white, middle class frame of care and concern.

Miss Jessie's family has owned this small square of property under a giant coastal sky for more than seventy-five years. She grew up here surrounded by relatives. Her grandmother, Lillie Mae, supported her. "She worked seven days a week," said Miss Jessie. "She was a master presser at the dry cleaners. I don't know if anybody knows how hot it is in there. I didn't know until I was in college. I came home to see her, and I stepped back there and thought, 'Oh my god, my [grand]mama has been working all these years to support me in this kind of heat.'"

"On Sunday mornings," Miss Jessie continued, "she ran the nursery at the Nazarene Church, then she would come back to our church, Second Baptist, and I was already there waiting. Our lives were centered around our church. If there was anything to be done at the church, dinners, or other gatherings, she and I were a two-woman restaurant. If we were raising money, she was gonna win. I learned to be a winner from her. I learned to work."

Miss Jessie draws on this ancestral strength and her faith in God to hold her ground in a community devastated by years of disinvestment and neglect. "The city never cared about us," said Manning. "They never did any improvements, no sidewalks, nothing." The one street the city of Freeport widened and maintained was Fifth, slated to be an entryway to the expanded Port.

"You've seen the condition of the streets," agreed Miss Jessie, nodding toward the window. "We pay the same taxes as everyone else. We don't get a discount. There is one piece of curb the city put right in front of the house, decades ago. I carved my name in it."

Landowners sold out to the Port or abandoned their property over the years, in part because the city would not issue permits for improvements. Ten years ago, after refusing to give Manning permission to improve, the city condemned his grandmother's house on East Second Street and forced him to tear it down, but he's hanging onto the land.

"For a long, long time, people here would not sell," said Manning. Then the Port started targeting the churches with million-dollar buyout deals. "Once the churches started to sell, that was it. That broke the community."

"And then people came out here door-to-door and started telling families they should sell out, and if they didn't the Port would take the land anyway," added Jessie. "And that made people fearful."

Three hundred and fifty-four houses once lined the East End, from Second to Eighth Streets, and at least half a dozen churches. Now one church and about forty property owners remain, and a few scattered homes. The lots on either side of Miss Jessie's sit vacant—both once belonged to cousins. "The day they came and knocked down my cousin's house next door, that was when the reality set in," said Jessie. "I just lost it that day."

The cousins' houses still exist on Google Maps, one light blue with white trim, and one beige, with palm trees out front—like starlight, the image is a reflection of something no longer there. In real life, Miss Jessie's place sits alone, a brick house on a nearly empty street. "My mother said if she could work for white people in brick houses, she could have one of her own," said Miss Jessie. She told me that when she was a kid other girls bullied her for having such a fancy place.

Manning is plaintiff in a lawsuit against Port Freeport for violating civil rights laws by coercing landowners to sell their property at unfair terms. Brought by attorney Amy Dinn at Lone Star Legal Aid, the suit doesn't ask for money. Instead, it seeks a fair and transparent compensation process for landowners. "Manning is the plaintiff, but all of us in the East End support this," said Jessie, who helps lead a group called East End Family.

"This is a really important case," Amy, the attorney, told me over the phone, before I visited Freeport. "These port expansions are happening everywhere, and that threatens communities. They get sacrificed in the name of profits."

Back in Freeport, Miss Jessie took us out front to show us her name, barely visible, carved into the piece of curb unconnected to any sidewalk. "This doesn't mean anything to them," she said, waving her hand toward the Port. "But I know how hard my grandmother worked for this house, to leave something for me. This place is my heritage and my legacy. That is being taken from me."

What racism stole from Miss Jessie and the other East End residents: the beating heart of this place, which infused generations with its lovingly nurtured life, and its wealth, a word that carries "well-being" in its earliest sense.

JOY

From Freeport, Aster, Cobalt, and I drove an hour north to my friend JP's in Houston. We pulled up at the modest house with its brick façade and chain-link fence. JP brought us to the back, shaded from the heat and overflowing with plants. Aster and JP's six-year-old daughter, Elena, played with a skinny neighborhood stray cat, pure white, that they had named Flaca.

JP is a seventh-generation Texan descended from German immigrants. They grew up outside the state but wanted to reconnect to their family past, so they moved back to Houston twenty years ago, to the working-class neighborhood where their dad was raised, in an area also called the East End.

JP showed us their garden with its many food plants, some we could never grow in Portland—banana, papaya, dragonfruit—and others that were more familiar—rosemary, tomatoes, kale. Also a maple tree, and a vitex, Louisiana irises, succulents of various kinds, impatiens. Some of JP's plants hold deep significance: a begonia their paternal grandmother grew and shared cuttings of among the family, and society garlic, a smell that JP remembers from childhood in their grandmother's yard, or permeating the city after a rain.

We sat amidst the greenery, and JP said this to me: "Please don't be one of those people who comes from outside and writes about nothing but what a hellscape this is. We have lives here that are full. We have beauty in our lives."

The next morning, JP took us to see the exhibition *citysinging* at the nearby Lawndale Art Center, where they had been in residence for the past year along with two other artists. JP is a writer, artist, translator, and language justice advocate. They describe what they do as "undisciplinary" work. What I experience about JP is that they do everything with deep intention. Their life is an extension of their art, and vice versa. The two flow together.

Another friend, the poet, artist, and translator Jen Hofer, was in town from Los Angeles. Together we wandered through the poem objects JP had installed for the piece *Unsettlements,* focused on sites around Houston of family memory, trauma, and violence.

JP conducted rituals at each of the sites and gathered sounds, images, and objects to evoke these hidden histories and the pain they still carry: whispery lines of Spanish moss trailing across the wall; a set of mirrored shelves stacked with mysterious items; a sort of mobile hung with caution tape, Styrofoam, and wire. Rather than telling the stories of these sites, JP focused on creating an experience of them, what they called a "non-telling." *Not*

how the scar came to be, but how it hurt...

Each object asked for engagement, an attention to its collections and their arrangements. After awhile, though, I could no longer ignore the sounds of Aster and Elena shouting and playing on the other side of the wall. I found Aster furiously pedaling an exercise bike, which was making a projection on the wall whirl faster and faster; it was a brightly colored snake made of what looked like plastic cups, spinning in the shape of infinity. I read the title of the piece: *Ouroboros, for Kekulé.*

The artist, Julia Barbosa Landois, was born and raised in Texas. She told me later in a phone conversation that she had spent some years working at an art gallery in San Antonio. "I watched artists come and construct these elaborate installations, and afterward most of it would go in the trash," she said. "I decided I didn't want to participate in that kind of waste." She started looking around for things to repurpose in her art. She had lots of unused yogurt containers in her studio. Then, in her research, she came across Kekulé and the ouroboros. So she started connecting the containers into the shape of a snake.

In 2017, Julia moved to Houston. A few weeks later Hurricane Harvey flooded her neighborhood and left her trapped for days with her neighbors. "Harvey changed everything for me," she said. “It gave the work more urgency.” In addition to the yogurt containers, Julia also had brightly colored silkscreen prints lying around that hadn’t turned out as she’d wanted. She started cutting the screen prints into the shape of the yogurt containers, transforming the sculpture into a two-dimensional piece that she could animate, evoking Linus Pauling’s resonating benzene ring, and echoing the first trademark symbol for plastic: infinity.

In March 2019, a few months before the show opened, chemical storage tanks owned by Intercontinental Terminals Company caught fire, sending a plume of black smoke across Houston. The tanks held byproducts of plastic manufacturing and petrochemicals for producing gasoline.

The disaster caused benzene levels in the air to spike and prompted city officials to order people to “shelter in place.” The notice directed nearby residents to close all doors and windows, turn off fans and furnaces, and seal cracks and holes with wet towels and sheets. It contained the following advice: “If you have trouble breathing, contact 9-1-1 and cover your nose and mouth with a damp washcloth, then take slow, shallow breaths and try to stay calm.”

Julia called another piece in the show *A Flood of Feelings*. On Ebay, she found reproductions of the painting *The Entry of Animals into Noah’s Ark* by the Flemish painter Jan Breughel the Elder, made in 1613. On top of the painting, Julia screen printed the words “oh ______ it’s two minutes to midnight.” The phrase refers to the Doomsday Clock, created by the Bulletin of the Atomic Scientists in 1947 to warn of impending nuclear disaster.

In 2018, the board of scientific and expert advisers moved the time on the Doomsday Clock to two minutes to midnight, in recognition of the existential threats of climate change and nuclear technologies. The last time the clock had been set that close to midnight was 1953, after the U.S. tested Teller and Ulam’s Mike device, and the Soviet Union followed with its own hydrogen bomb explosion six months later.

Julia invited attendees of the exhibit's opening to participate in the piece. They could choose from a series of words and phrases to fill in the blank, including "oh #*@?!" "oh God," and "oh well." She told me that some attendees selected a number of different phrases, layered on top of one another. "I felt like this was a community ritual, a way to process the catastrophe of Harvey," said Julia.

I asked her about humor and play in her work: the bright colors, the bike-powered snake. "It's a way of holding your hand," she told me, "of getting you to stay here with me and look at these hard things. Besides, why would we want to preserve the world if not to experience humor together, and joy?"

Joy. It was for joy, or the memory of joy, that the artist Robert Hodge installed his piece for this exhibition, called *Leroy's*. Robert, a native of Houston's historically black Third Ward, recreated a 1990s-era Blues club inside the gallery. "Blues music is about pain and escape," Robert told me. "But what I discovered in the project is that these clubs were really about the community. It was about hanging out over food, a cold beer after a long day, taking your lady out. The musicians did it for love. They loved to sing and perform, and people from the neighborhood came out just to enjoy it, enjoy being together."

The clubs are now mostly gone. Robert remembers ads for them as a child, but he never visited one. For the installation, he drew from the collection of a woman named Melissa Noble, who has been gathering memorabilia from Houston's clubs for thirty years. "I asked her why she collected so much, but she said she didn't know," said Robert. "I think it was because she found community there. Even though she was white, they accepted her, they were kind to her. She had posters, albums, dominoes sets. She even saved the plastic cups for the chandeliers."

"Chandeliers?" I asked.

"You know they didn't have any money, so they used to make chandeliers out of plastic cups," he said. "They would wrap Christmas lights around them. The clear cups have aged so they're yellowish and then the blue lights come through. It's beautiful. It's just an angelic look."

The gallery club, Leroy's, which is Robert's middle name, had folding chairs and tables with white cloths and fake flowers, posters of musical acts on the wall, and portraits of Houston Blues artists. Silver fringe hung behind the stage, and a wooden sign read: "SOULFOOD KITCHEN B-B-CUE."

I didn't notice the plastic cup chandelier at the time, but I saw it afterward in the catalog photo, hung in a corner, a sphere of plastic cups glowing blue. I remember the silence of the little room, the absence it evoked. It almost ached for music, for life.

Robert brought singers to the space. Miss Trudy Lynn and Odell Gray, who performed in the original clubs, as well as a younger act called The Peterson Brothers. He's making a documentary about the Blues scene in Houston, capturing a history that might otherwise disappear completely. He also just bought the house where he grew up in the Third Ward. After studying art in Atlanta and New York, he wanted to come back to his community. Like JP, Robert's life is an extension of his work, and it draws sustenance from its roots. He told me, "I preserve history for the sake of the present, and to point the way to the future."

REMEMBER

I returned to Freeport in November 2019, flying solo to Houston (1,056 pounds of carbon dioxide) and renting a car. I met Miss Jessie at the Texas Roadhouse in the Brazos Mall in Lake Jackson. While I waited for her in the lobby, I studied the Tribute Wall with its photos of local soldiers who had died in Iraq or Afghanistan. There were four of them, young white men, all nineteen or twenty years old.

On the opposite wall was a painting of an octopus emerging from the water and seizing a boat, one of its tentacles entwined in fishing line, one inscrutable eye visible. Only the boat engine remained, emblazoned with the TEXAS ROADHOUSE logo. A cartoon pig wearing sunglasses and holding a fishing reel flew through the air, a terrified look on its face, as the octopus engulfed the boat. An apocalyptic sky stained the water red. Over it all loomed industrial towers in black silhouette, and the red Dow logo.

When she arrived at the restaurant, Miss Jessie seemed worn out. She had spent the day contacting pest control companies, battling rats driven to her house by the Port knocking down buildings around her. Usually when we talked, I listened more than spoke. I wanted to understand what my service would be, how I could serve the agenda of Miss Jessie and her community—or, if you ask Miss Jessie, serve God's agenda through her.

She and Melanie and Manning thought, in a vague way, that maybe a book could help them by bringing attention to the struggle of this small town in the shadow of Houston. I had zero faith that a book could do anything, but book writing, and my paid work for an environmental group, involved skills that might help: writing, research, fundraising, project managing, and a few connections with journalists and lawyers.

Miss Jessie consented to my involvement. She saw God's hand in it, but on this night, she was tired. We ordered ribs and Miss Jessie bowed her head to pray. Then she complained to the server about the dinner rolls being undercooked and sent them back. Then we talked for three hours.

A few months earlier, in September, the newly appointed U.S. magistrate judge in Galveston dismissed Manning's case against Port Freeport. In one of his first decisions as a judge, Andrew M. Edison said that since the lawsuit did not explicitly state that Manning is African American, his race was unclear. He also concluded that the case did not prove the Port was intentionally discriminating. A month later the same judge also dismissed Manning's related civil rights violation case against the Army Corps of Engineers, which is providing more than half of the funding for the Port's expansion. Lone Star Legal Aid was working to appeal the case in the Fifth Circuit, which covers Texas, Louisiana, and Mississippi, and is known for being friendly to industry.

On October 10, 2019, the Port Freeport commissioners voted unanimously to move ahead with eminent domain, empowering themselves to take the remaining East End properties in exchange for compensation.

After the vote, the Port held a meeting for area ministers that Miss Jessie found out about. "They didn't know that I am a minister, so they didn't expect me, but I came!" she said. "And I was the only one who came, so I got to speak my mind."

Miss Jessie said that the Port CEO, Phyllis Saathoff, who makes $212,000 a year, said that she understood Miss Jessie because her own father had worked in the cotton fields. "'You don't understand me,' I told her," said Miss Jessie. "'You're not Black. I'm going to be Black until I die.'"

: :

We left the Roadhouse and I drove, at Miss Jessie's suggestion, about two hundred yards to the Marriott across the parking lot. I knew exactly what I would get: the anonymous comfort of a dark room, crisp linens, fluffy towels, feather pillows. I wanted that.

The next morning, I remembered the story about enslaved Black people buried near the mall. I decided to visit the site. I asked about it at the front desk. The young white woman working told me she knew where it was because she had played in that area as a child. She sent me to the opposite corner of the parking lot, to a green space bordering Oyster Creek, which runs behind the mall. It was a beautiful, sparkling fall day. I walked all around, but I saw no markers of any kind. So I went back to my rental, a giant silver SUV I'd been assigned after asking for a compact, and there, in front of my parking space, I saw the plaques. Workers had uncovered the remains when they were building the Marriott where I'd slept.

I learned later that the mall owners knew about the overgrown, unmarked graveyard called Mt. Zion Cemetery—they just planned to build around it. But the construction crew, working with archaeologists, found the remains of three humans, and maybe one cow, outside the graveyard's known borders in 2014. The mall sits less than a mile from the Abner Jackson plantation house, on land that was part of his sugar mill. Harold Gaul, a member of the Texas Historical Cemetery Guardianship Association, told the local newspaper that the group has the names of seventy people buried in the area, and more likely lie beneath the soil.

"Even though the mall has spent a significant amount of money on plans and engineering, we have now completely stopped," the mall manager, Patty Sayes, told the local paper. The accompanying photo shows Sayes, a white woman wearing a black dress, pearls, and tennis shoes, pointing down at the bare earth. Gaul stands off to the side, a Black man dressed in some sort of uniform, hands behind his back, looking at the camera.

A year after the discovery, the city and mall officials held what *The Facts* described as a dedication of the cemetery as an official Texas historical site. The paper quotes Anthony Bonner, one of about a dozen people in attendance with ancestors buried there. "This is not a dedication," said Bonner. "It's a realization that this area was basically built by these people."

I got out of the truck and walked toward two granite markers resembling gravestones. They sat in a grassy area beneath a curved and spreading live oak. Cars roared past on the expressway toward the plants. The marker on the left read:

COMMEMORATING THE SOULS OF THE AFRICAN

AMERICANS WHO WERE

INTERRED AT MT. ZION CEMETERY

DEDICATED THIS 12TH DAY

OF SEPTEMBER, 2015

LAKE JACKSON, TEXAS

BRAZORIA COUNTY

The one on the right, in smaller type:

COMEMMORATING THE LEGACY, INTEGRITY

AND RESILIENCE OF THE AFRICAN AMERICANS

INTERRED AT THIS FINAL RESTING PLACE.

CEMETERY AT ABNER JACKSON PLANTATION

DEDICATED THIS 12TH DAY

OF SEPTEMBER 2015

LAKE JACKSON, TEXAS

BRAZORIA COUNTY

Except for the use of the word "plantation," neither records anything about slavery.

I admit to spending time obsessing about the period following the word "place." It is the only one on either plaque. It makes the final resting place of these humans seem very final. It also makes the phrase "CEMETERY AT ABNER JACKSON PLANTATION" seem to float endlessly, inserting itself into the present, the future.

On the way back to the truck, I noticed that someone had planted thirteen sycamores along the edge of the parking lot, probably when they'd finished the Marriott. Someone had stabilized the young trees by wrapping their trunks with metal wires connected to stakes. No one had ever removed the wires, and the trees had grown around them, each one scarred and gashed by the metal. I knew that the wounds would weaken the trees and open them to disease, most likely killing them.

As I drove out to meet Melanie for breakfast at Denny's, I noted that on the north side of the lot, facing the expressway, someone had more recently planted young trees and tied them, once again, with wire.

: :

Miss Jessie speaks often about seeing it as her purpose to "break the generational curse." I asked her once to tell me more about this. "The older generations didn't talk about their lives," she said. "But if you don't know what happened to your mother and your grandmother, and the one before her, and the one before her, how can you change the future? I decided to supply my children with all the information I could so that they could make different decisions. We need to know what happened to us."

Remember

JOY

The Denny's in Lake Jackson was packed with post-church Sunday brunchers. While we waited for a table, Melanie showed me a flow chart outlining the eminent domain process. Melanie had scrawled notes across it during a recent meeting for residents held by Lone Star Legal Aid. Melanie's notes had words like "appraisal," "judge," "challenge," "eviction," "attorney." At the bottom, below a red box labeled "Final Judgment," Melanie wrote, "Port can take your property."

Melanie told me some more about the town. That Dow had donated the land for the Lake Jackson Cancer Center. That she didn't like going to the library because she found it too depressing to see all the permit applications companies have filed for new construction. "Why would our leaders just give away everything to industry and the Port?" she said. "We have a beautiful city here. We have a beach." She told me Freeport is surrounded by levees and that it did not flood from Harvey. "This could be a good place for people."

After breakfast, I drove around the plants, awed all over again by the sprawling landscape of metal towers spurting steam and flame. I parked in a lot beside a canal where I could see Dow's eight cracking furnaces for making polyethylene. I listened to the plant; it made a hissing sound, the sound, I guessed, of high-pressure chemicals moving through miles of pipes. It was a giant sound, loud, but not ear-splitting, just...massive. It filled the air from all directions, drowning out everything else, except, I realized, the wind. A cool, salt breeze was rippling the grasses all around me. I stood only a few miles from the Gulf Coast. As Melanie said, "We have a beach." I turned my face to the scent and took it in. In honor of beauty, in honor of joy, I decided to follow it.

I drove south through town, past the plants, past the port, over the ship channel, and turned left at a turquoise sign that said Quintana, which means "country house." The huge blue sky seemed to press the land down flat. On the right, among scattered palm trees, pastel colored houses rose on stilts. On the left marched a line of high voltage wires and, behind them, behind an earthen berm and a razorwire fence, sat a series of plants for liquefying natural gas. The plants are called trains. They are long and narrow and silver. In aerial photos they look more like a fleet of enormous spaceships landed on the coast.

I turned right onto Eighth Street at the Quintana City Park, the gleaming new plant facilities visible across the street. I drove past the Quintana Neotropical Bird Preserve, a small patch of land that looked more overgrown than the surrounding landscape, and I continued until the road dead-ended in sand. The plant was easier to see from the beach, its regimented, silver towers incongruous against the sweep of sky and sand. Two men in cowboy hats, one Black and one white, ambled past on horseback. Kids played in the sparkling water. I sat in the sand and let the sun warm me.

Another way companies are making money off the cheap natural gas freed by fracking is to cool it to minus 260 degrees, which turns it into a liquid with a much smaller volume than gas, and ship it to other countries. It's called liquid natural gas, or LNG. Freeport LNG, once a small natural gas importing facility, saw that coming and wanted in, part of a wave of new LNG plants across the U.S., mostly on the Gulf. In 2011, the company announced it would seek approval to build a new $14 billion export facility.

The project divided the small town. The Sierra Club jumped in. They filed suit, arguing the federal government had to consider the damage to the climate these plants would cause for generations, at a time when the United Nations' panel of scientists warned that the world had only a few years left to drastically reduce pollution. But federal courts rejected a half dozen lawsuits against LNG plants, including Freeport's, which instead received approval to expand.

"Sierra Club considers this a failure, but there are a lot of positives for us," Melanie told me. "We slowed construction down for a year-and-a-half and put pressure on them to do it right. The government put ninety conditions into their permits. No one had ever built an LNG plant on a dredged fill-site before. They had to go back to the drawing board and dig the pilings deeper than they ever have and use more concrete. They claim that the plant will stop storm surge and hold up to a hurricane. There are conditions on conducting air monitoring, how high of a [protective] wall they need to build, how close the [gas] flare can be to homes."

I hadn't planned to go to Quintana. I just drove south following the smell of the Gulf. I sat there, watching the blue water lap gently over the sand. The Gulf showed no visible marks or scars from the 200 million gallons of oil that BP spilled almost ten years ago. I knew these waters teemed with fish that fed people and gave them jobs. I knew that most of the world's species of sea turtles swam these waters, along with dolphins, whale sharks, and vampire squid with their glowing clouds of mucus. I knew that despite all the insults and injuries inflicted on this water, it continued to nurture these wonders.

Later, I thought of something else JP had pointed me to, the words of the botanist Dr. Robin Wall Kimmerer, who is of European and Anishinaabe ancestry, and an enrolled member of the Citizen Potawatomi Nation. She wrote in her book *Braiding Sweetgrass*:

> Even a wounded world is feeding us. Even a wounded world holds us, giving us moments of wonder and joy. I choose joy over despair...because joy is what the earth gives me daily and I must return the gift.

I thought of what JP had written me: "I know the story here in these lands is not all heaviness, loss, and sadness...if those were the only stories we would all be dead already...There is a joy to the struggle. A joy to surviving."

And I remembered the words of Melanie, that day at Miss Jessie's: "We are building strength in numbers. People are paying attention. So we continue to speak up, every way we can. I'm not giving up."

And Manning, the other East End resident I met with that day, who looked down at my digital recorder, and back into my eyes. "I'm Manning Rollerson," he said. "I'm not afraid. I know that whatever happens, I did the best I could for this community."

DIS-COURAGE

Errol Summerlin drove his twelve-year-old Cadillac sedan slowly around a two-mile scar of dirt cut into the farmland in Portland, Texas, a small city just outside Corpus Christi. A security guard in a white pickup truck crept along behind us for a while, then passed. "They know my car," said Errol. "He probably called it in, and they told him, 'Oh, it's just that Summerlin guy again.'"

Inside the expanse of fence, surrounded by an earthen berm, rose a few construction cranes and a beige, blocky building with the sign GULF COAST GROWTH VENTURES, and a blue and green logo that looked like eight interlocking benzene rings. Below it were the names ExxonMobil and, in Arabic and English, SABIC, which stands for Saudi Arabian Basic Industries Corporation, the largest public company in the Middle East.

On the other side of the road sat some modest houses, and an expanse of sorghum and cotton fields, with massive white wind-turbines rising out of them, an extra source of income for farmers. Immediately to the northeast lay the town of Gregory, a low-income, largely Latinx community.

The site will soon be home to the world's largest ethane cracker plastic plant, exceeding the Dow plant in Freeport and sprawling across an area twice the size of the town of Gregory itself.

Exxon and SABIC inked the deal for this plant in 2017, when President Trump traveled to Saudi Arabia with former Secretary of State Rex Tillerson (also the former CEO of Exxon) and the current Exxon CEO, Darren Woods.

The companies paved the way for the project long before the official handshake. "In the fall of 2016, it was all of a sudden on the school board's agenda to give 'Project Yosemite' tax abatements," said Errol. "We had no idea what that was."

Companies often use code names for big new development projects, ostensibly to keep competitors in the dark; but this also keeps communities in the dark. Sometimes even local officials don't know the companies to whom they're handing out tax breaks.

Errol has a trim grey beard, dark brown eyes, and a deep voice, made gravelly from smoking. Before he retired, he spent his career as a legal aid lawyer helping people without resources navigate the legal system. He has lived for thirty-five years in Portland, a place like Freeport—and my Portland in Oregon—linked by water to global commodity flows. The word port comes from an ancient root that means "to pass through or over." Where goods transit, people also settle, seeking refuge and shelter—two later meanings of the word port.

Errol has leapt into other struggles here against development: Corpus Christi's Refinery Row, across the bay from Portland, has one of the largest concentrations of oil refineries in the nation. When Errol got wind of the mysterious project close to his house, he went into action.

Today, July 4, he was taking a break to show Cobalt, Aster, and me around. From the plant, Errol headed to Sunset Lake Park, on a strip of land that sticks out into Corpus Christi Bay. He wanted us to see what gives him joy in this place. He pulled to the side of the road and we climbed out into the hundred degree heat. He spun his baseball cap around backwards and held a pair of binoculars up to his eyes, then handed them to Aster so she could see. Improbably pink roseate spoonbills dangled from the low foliage. Tricolored and blue herons stalked the water for prey, alongside glowing white egrets. Stately brown pelicans beat their wings across the surface.

To the north is Aransas National Wildlife Refuge, where endangered whooping cranes migrate in winter. In an average year, 700 billion gallons of stormwater from the plant's equipment and processing will run to Copano Bay near the refuge, Errol told us.

We'd arrived in Corpus Christi the night before, just in time for a meeting of the Coastal Alliance to Protect our Environment (CAPE). Errol welcomed Cobalt, Aster, and me warmly with hugs and ushered us in. About two dozen people had gathered at the Citrus Bayfront Bistro on a sultry Wednesday evening. They were fairly diverse in age and ethnicity, representing a variety of local citizens groups, including For the Greater Good, Port Aransas Conservancy, and the Islander Green Team, as well as local chapters of national groups, such as Surfrider and Sierra Club.

Errol and a few others in the room had been resisting and delaying the Exxon-SABIC plant for more than two years, first contesting the $1.1 billion in tax breaks it received, and then challenging the pollution permits handed out by the Texas Commission on Environmental Quality. A few weeks before I arrived, the plant had just received its final air pollution permit, making it a fait accompli. It was hardly a topic at this meeting.

Errol opened the proceedings by describing the drive he had just taken with his wife from New Mexico on Highway 285. We had driven part of this road also; it's the one highway from Los Alamos. It travels south the length of New Mexico and into Texas through the Permian Basin, the bed of an ancient inland sea.

The sediments at the bottom of the sea compressed over millions of years into shale rock, layered with dead sea plants and animals that transmuted over time into liquid oil and gas. The Permian holds one of the world's thickest shale formations, extending more than a thousand feet in some places and harboring the planet's largest known concentration of hydrocarbons, molecules consisting only of hydrogen and carbon that make up fossil fuels.

Oil companies have been drilling into the Permian for decades. It supplied much of the fuel for the military in World War II, but production dropped in this century and many companies left, until fracking brought the Permian back to life.

Hydraulic fracturing—fracking—involves pumping water, sand, and chemicals (a proprietary mixture companies do not disclose) into rock at high pressure to blast open the shale and free trapped oil and gas. Before fracking, the industry considered much of the remaining oil in the Permian "tight," too locked inside the rock to reach. Since fracking became widespread, the amount of oil extracted from the Permian has quadrupled, and it is now the world's second largest producing oilfield behind only Saudi Arabia. The most active driller: ExxonMobil.

The two-lane portion of 285 that runs through the Permian has become known as the "Death Highway" because of collisions with the thousands of trucks moving equipment to and from the oilfields. "Stay alive on 285" is the mantra.

"It's flat, so you can see for miles," said Errol, "and what we saw was a nightmare, creeping along at fifteen to twenty miles an hour with hundreds of trucks, past hundreds of drilling sites. I saw trucks carrying pipes for new pipelines, pipelines that are all headed to us. They're all headed here."

The sense of threat in the room was visceral: the fear that the planet's largest concentration of oil and gas was about to come dumping down on their heads. Three different companies are building pipelines from the Permian to the Port of Corpus Christi, which aims to grow from third place to the largest oil export hub in the U.S.

"We need an overview of all the projects underway," announced Errol. He stood beside a map of the region, circling with a pen as another group member listed off new developments, eighteen in all, ranging from projects to dredge the port and ship channel to new fractionaters, which turn gas into usable products; an expanded liquid natural gas plant; offshore oil terminals; another new plastic plant; something called Project Dynamo; and something called Project Falcon.

"And last," said Errol, rounding out his list with another offshore oil terminal. "Second to last," interjected a participant who had heard of another project. "Actually, there are a few more," said another. Others piped up with more, until the list of development projects grew to nearly two dozen.

"All we can do is be watch dogs," said Errol. "We have to watch everything that's going on, try to slow it all down, contest their permits, and bring awareness to what is happening here."

The next day, after driving me around town, Errol headed to his house, a 1970s white brick ranch. His wife, Anna Maria, who spent her career as an educator, was out front watering. Inside was crowded with three decades of art, memorabilia, and photos of their two sons and their grandchildren, along with, Errol estimated when I asked, about seventy potted plants.

He took me to the screened-in porch out back where he and Anna Maria watch migrating birds come to the many feeders they have put out. Then it was off to what he calls the "war room," a spare bedroom decorated in green. I sat on the coverlet while he pulled up maps on his computer showing me the areas he had just driven me around: the plant site, the outfall pipes, the neighborhoods at risk.

He navigated to the website for Gulf Coast Growth Ventures and had me watch a promotional video about the project. Using computer animation and a soothing soundtrack, the video depicts a cotton field transforming magically into a chemical plant, with each of the components clicking cleanly into place. It notes that the companies will build giant sections of the plant elsewhere—the location is never indicated, though Errol presumes China—and ship them to the Texas coast.

The pieces of the plant, some as long as a football field and as tall as the Statue of Liberty, will travel from the coast to the site on massive moving machines, the kind that carried the space shuttle out to launch. The partners built a five-mile "heavy haul" road through the city for this purpose.

The animation brings home the massive scale of the undertaking. A later promotional video compares the ambition of the project to the moon landing. To make it explicit, the companies held a kickoff event in September 2019 at NASA's Johnson Space Center in Houston with former astronauts as guests, and a person in a spacesuit.

Presidents, astronauts, kings: It's the kind of project that only global companies with vast wealth could build, the wealth of empires.

Back at the CAPE meeting, Errol put his final circle on the now-crowded map and turned to the room. "It's daunting, what we're facing," he said. "Is anyone discouraged?"

The word discourage comes from the Old French word for heart and the prefix *dis*—to part or split. Discourage: to cut one off from one's own heart.

"Is anyone discouraged?"

It was a question no one answered.

The CAPE meeting lasted two hours. It was a mostly subdued, somber gathering, but at one moment everyone broke into applause: when Errol mentioned the fact that, ninety miles up the coast at Point Comfort, Diane Wilson had won her lawsuit against Formosa Plastics. She sued them for contaminating the waters where she and generations of her family worked as commercial fishers and shrimpers.

Diane's was a three-decade battle to draw the world's attention to the pollution dumped into Lavaca Bay and the surrounding creeks and wetlands on the central Texas coast. It included a series of hunger strikes and, in 1994, Diane's attempt to sink her forty-two-foot shrimp boat, Seabee, near the plant's wastewater pipes as a "permanent monument to the suffering of the bay."

People willing to join Diane Wilson in the fight came and went over the years, but not many wanted to take on one of the region's richest companies and top employers. Ten years ago, a former wastewater operator at Formosa contacted Diane about the nurdles spewing out of the company's pipes into the bay. Once released into the water, the plastic pellets pick up toxins like pesticides and heavy metals. Studies have found that nurdles have concentrations of toxins up to ten million times higher than surrounding seawater. The science is still new, but when shrimp, crabs, fish, and other sea life eat these little toxic pills, it seems possible that the poison gets into the food chain, and then into humans.

Diane spent several years contacting regulators and urging action against Formosa. The Texas Commission on Environmental Quality and the U.S. Environmental Protection Agency fined the company a few times for violations—$13 million here, a few million there—but the nominal penalties didn't change anything at the billion-dollar company, run by one of Taiwan's richest families.

In 2016, Diane and a few others, including former Formosa employees, created the San Antonio Bay Estuarine Waterkeepers. Several times a week they go out on foot and in boats to collect nurdles from Cox Creek and surrounding areas. They gather samples, and label them with the date, time, and location. "If we were trying to pick up everything we could find, there's no way we'd ever finish," Diane told a reporter from *Texas Monthly*.

The advantage of plastic waste is that it's visible. Over three years, the group collected an estimated 30 million nurdles, all stored in Diane's barn. The group sued Formosa for violating the Clean Water Act, and on the first day of the trial in March 2019, they showed up at the courthouse with thirty plastic bins full of nurdles. Security officers would not let the bins into the courtroom, so the judge ordered them kept in the basement where he and the lawyers could examine them.

Thomas McGarity, a law professor at the University of Texas at Austin, dismissed this kind of citizen-collected evidence in an interview with *The Texas Tribune*. "[It] is kind of a romantic idea," he said, but "there are all sorts of rules about preserving data that a nonexpert can run afoul of."

Judge Kenneth Hoyt disagreed. He called Formosa a "serial offender" and its violations "enormous," considering the evidence provided by Diane Wilson and the others.

Amy Johnson with Texas Rio Grande Legal Aid, who filed the case on behalf of the group, called it an important precedent for citizens looking to hold corporations to account when governments fail to do so. She told *Texas Monthly* that other clients are following Diane's example.

The plaintiffs against Formosa asked for a fine of $166 million, the maximum under the law. In December 2019, Judge Hoyt approved a $50 million settlement, the largest in U.S. history involving a private citizen's lawsuit against an industrial polluter. The money will go into a trust fund to clean up waterways and beaches. The settlement also requires Formosa to discharge no more plastic pollution.

Even after they won, Diane's group still went out collecting nurdles. She has been burned by Formosa agreeing and then failing to reduce pollution before, so she's not letting up until she sees changes on the water. "As long as it takes," she told *The Texas Observer*, "we'll be here."

Meanwhile, Formosa in 2020 received permits from the state of Louisiana to begin constructing a $9.4 billion chemical complex to make polytheylene, polypropylene, and other plastic products. The plant is located in the Fifth District of St. James Parish, Louisiana, on part of a stretch of the Mississippi River between New Orleans and Baton Rouge already so burdened by pollution it is called Cancer Alley. Once completed it will be, once again, one of the largest plastic factories on the planet.

The company is moving forward despite fierce and organized resistance from the area's residents, including many majority Black communities. "These are lives. These are human beings," said Sharon Lavigne, founder of Rise St. James, a faith-based group of local residents. "People are suffering and we're asking for help," she said in a statement, "but it seems like wealthy people and those in charge don't care."

The state of Louisiana gave Formosa $1.5 billion in tax breaks for the complex. The company codenamed it "The Sunshine Project."

HERE

Much of the discussion at the CAPE meeting in Corpus Christi centered on desalination plants. The city is planning to go into debt to build the state's first facilities to suck up seawater from the Gulf, press it through high-tech membranes, and collect the salt-free water.

The official line is that the plants will help ensure a water supply for the city in times of drought, but industrial expansion, or what the city calls "high-volume users," is fueling the drive for desalination. "The Exxon plant alone will use up to 25 million gallons of water a day," said Errol. "That's the equivalent of all the residents of San Patricio County," where the plant is located. "The city has promised industries water, and now they have to provide it."

The $222 million loan for the first plant will come from the Texas Water Development Board. The Board consists of three individuals appointed by the governor. One is Kathleen Jackson, a former PR executive for Exxon.

Much of the world's desalination capacity exists in three places: Saudi Arabia, the United Arab Emirates, and Qatar. The problem with this technology: It requires massive amounts of energy, and pollution, to filter the seawater. For every gallon of fresh water derived, at least one more gallon of warm, super salty water, often contaminated with chemicals, goes back into the ocean. The plants in Corpus Christi will dump 131 million gallons of salty, chemical-laden water every day into Corpus Christi Bay, according to Texas Campaign for the Environment.

The CAPE discussion was rational and intellectual. It touched on technical aspects of "desal" plants, possible locations, and who will pay for them. Then a woman with a long black braid, wearing a t-shirt that said, "Check your privilege," stood to speak.

"I just want to interject here," she said. "The legal strategies are important. The technical aspects are important. But most of us don't have that expertise. We don't have material power, either. What we do have is people power. And that's what we need. We need to get out and meet people where they are and talk to them about what is happening here in their community."

I learned that this was Dr. Isabel Araiza, a sociologist and associate professor at Texas A&M University in Corpus Christi. We spoke on the phone a few weeks later.

Isabel was born and raised in Corpus and has deep roots in Texas. She grew up near the county jail and the refineries and was part of the first court-ordered integration of schools in the city, something she didn't realize until she was an adult. While researching the education system in the U.S., Araiza came across a footnote referencing a 1968 court case: Cisneros v. Corpus Christi Independent School District. The decision extended the Supreme Court's 1954 Brown v. Board of Education desegregation requirements to Mexican Americans.

Corpus Christi dragged its feet for years, only beginning desegregation in 1975. Araiza started kindergarten in 1978. She remembers waking up early and waiting in the darkness with her mom in front of the public library to board a bus that would take her across town to school. "I had no idea that was part of this historical moment," she said in a lecture at the local Del Mar College.

When Araiza finished high school, she attended Del Mar. "I didn't know anything about colleges or financial aid, and there was no one to advise me," she told me. "I only knew what was here." After college, she was teaching GED classes and stumbled onto the field of sociology. "It gave me a vocabulary to name what I saw, felt, and knew," she said, about the inequity affecting her community.

Isabel earned a Ph.D. at Boston College. "I realized I needed to come back home and share what I was learning," she said. She got involved in activism because of a series of crises in the city over potential industrial contaminants in the water supply.

"There was an emergency meeting. A whole bunch of people were there complaining about the impact on communities. We found each other and started strategizing," she said. The group eventually formed For the Greater Good, which Isabel described as "about a dozen ornery people."

"My philosophy is start where you are, and work with what you have," she said. They made a banner out of an old tablecloth, lettered t-shirts with duct tape that said "WATER," and made signs declaring, "Industry gets tax breaks, and we get DIRTY WATER."

"We started showing up at all the Council meetings," she said. "Officials were dismissive of us, but we knew the public was watching. That helped us to be brave and disruptive at news conferences that were really just PR events for the city. We got on the CBS national news, and officials finally felt forced to deal with us."

I asked her about the fact that people need jobs and economic development. That's the big selling point for industries. "Actually, if you look at the numbers from the Bureau of Labor and Statistics, industry jobs are much smaller than health and education," she said. "Those are areas where we really need people. We cannot get enough medical professionals to come here."

Isabel echoed the findings of other experts. In her book *Strangers in Their Own Land*, sociologist Arlie Hochschild points out that industries are highly automated and need a small number of well-trained operators. Hochschild's focus was Louisiana, but she found similar results. More jobs are in education, healthcare, construction, and other fields than industry.

Hochschild cites Dr. Paul Templet, a chemical physicist and retired professor at Louisiana State University, and former head of the Louisiana Department of Environmental Quality. He describes a downward spiral: Industries get tax breaks, which erode public services like schools and healthcare, degrading people's ability to get good jobs and earn an income. The companies then "give back" to communities, in the form of scholarships or support for environmental initiatives, and people become dependent on industries "not just for jobs, but for gifts," said Templet.

In Germanic languages, the word gift means "poison"—from the Greek *dosis* for dose, a giving.

The poisoned gift is familiar to Isabel. "Our schools are in crisis," she said. "My daughter had black gunk coming out of the water fountains in her high school. Mice run in the hallways. That's what tax breaks get you. Everybody knows our public services are in trouble, they just don't make the connection with industry because companies have done such a good job of promoting themselves as stewards.

"You've been here," said Isabel. "It's stunningly beautiful. I was in Costa Rica ten years ago. I would wake up every morning and be in awe of the beauty. Privileged people escape to those beautiful places, but if we just took care of what we have, we could have that right here."

TIRED

I was in the bathroom at the Best Western in Portland, Texas, when Christine Bennett called. I had been trying to reach her and other leaders of Mossville Environmental Action Now for several weeks. I hustled out of the bathroom and grabbed a notebook to write down our conversation, phone pressed between shoulder and ear.

Mossville is in Louisiana, about an hour from the Texas border, off Interstate 10. I have seen widely varying references to its founding year. It may date as far back as the 1790s, one of the first Black communities in Louisiana. It thrived as a haven for Black families through slavery, the Civil War, Reconstruction, and Jim Crow. Residents fished, hunted, grew food, and worked as domestic servants in nearby Lake Charles and in the timber industry. In World War II, refineries came to town, followed by chemical plants, taking advantage of the nearby oil and gas reserves, and a ship channel to the Gulf of Mexico.

Eventually fourteen plants encircled Mossville. A company named Condea Vista built a factory to make vinyl. Along with other factories in the region, Vista polluted the town's air and water with chemicals like dioxin, one of the most toxic substances known.

Christine and her husband, Delma, helped lead the fight against pollution in Mossville for decades. Along with other leaders, they formed Mossville Environmental Action Now (MEAN) in the 1980s. In 1998, MEAN and activists from Greenpeace convinced the federal government to conduct a study, which found that residents had dioxin levels triple that of the general U.S. population—some of the highest dioxin rates ever recorded.

The body can't break down dioxins, so they build up and cause harm for years—cancer, heart disease, diabetes, endometriosis, early menopause, reduced testosterone and thyroid hormones, skin, tooth, and nail abnormalities, damage to the immune system. Dioxin creates hormonal and genetic wreckage in developing fetuses, harming the brain and spinal cord—the central nervous system—affecting thinking, behavior, and growth. A separate 1998 study of Mossville residents found that 84 percent had some type of central nervous system disorder.

The federal scientists at the Centers for Disease Control and Prevention who found the dioxin levels could have located their source pretty easily. It requires identifying the chemical signature of dioxins emitted into the air by the chemical plants and comparing it with the dioxins in residents' bodies. They didn't do this, but the independent chemist Wilma Subra carried out this analysis for the community in 2007. In most cases, she found dioxins in the blood of people that were identical to dioxins from the plants. She submitted her findings to the federal government, but it gave no response.

There is good reason for this, Dr. David Ozonoff, professor of environmental health at Boston University, told the investigative news site *The Intercept*. Holding polluters accountable could jeopardize jobs and tax revenue that elected officials depend on. The accused company also could turn on government scientists and attack their credibility. If you don't do the analysis, "your only opponents are the activists in that given town—and they won't have many resources," Ozonoff told the reporter. "They can be tenacious, but they get tired after awhile."

In 2001, South Africa's largest company, SASOL, or South Africa Synthetic Oil Liquids, absorbed parts of Condea Vista and continued operations in Mossville. The apartheid government created SASOL in the 1950s to process oil out of domestic coal reserves, which helped the racist regime survive when nations refused to sell them oil. Turning coal into liquid fuel is an energy intensive and highly polluting process. The company's Secunda plant in South Africa is the largest single source of climate pollution on the planet.

In 2010, Mossville residents won a hearing before the Inter-American Commission on Human Rights, on charges that the U.S. Government had violated international law by failing to protect them from pollution. It was the first time the Commission agreed to such a hearing. The same year, CNN featured Mossville as one of the nation's most toxic towns.

In 2012, SASOL, now a private company, announced plans to build a $9 billion ethane cracker plant for plastics, again one of the largest in the world. SASOL offered to buy out most of the remaining residents of Mossville to make way for the expansion. The company funded an oral history project, a book, and an exhibit at a museum in Lake Charles to commemorate the community they erased.

A 2016 article in *Chemical and Engineering News* includes an interview with SASOL's VP for public affairs, Michael Hayes. The article describes him as "palpably frustrated" by negative press about the project. He rattles off the good things SASOL has done for the community.

"You aren't going to write that tired old Mossville story again, are you?" Hayes asked the reporter.

: :

When I picked up the phone in Corpus Christi, it became clear that Christine Bennett wasn't actually sure who I was. She said she saw my number several times in her phone and dialed it back. I explained that I was writing a book about plastic and asked if she would be willing to have me visit to talk to her and Delma, both leaders in the struggle against pollution in Mossville. She paused for a moment, then said, "You know, we've talked to so many people. We're talked out. We're tired."

"I was born and raised in Mossville," she continued. "This was my home. We just want a memorial put out front, we can't even get that.

"Eighty-five percent of our land now belongs to a foreign company. Can you believe that? They destroyed the homes of poor people and people of color, and nobody helped us fight. They are killing us, and all they want is a dollar. I buried seven family members from cancer this year. We are *dying*," she told me.

"I'm so sorry, Mrs. Bennett," I said.

"I appreciate that you feel sorry," she replied. "But sorry isn't going to bring back my dead. Sorry isn't going to bring back home.

"They came in here and destroyed our home, destroyed our community. A book might help someone some time in the future. It's not going to help us.

"There's no reason for us to go wild and get crazy with anger," she continued. "We just stay calm and trust in God. He's the only one who will help us."

She *was* calm. She didn't raise her voice or shout. But she was burning with anger. It crackled through the electrical signals connecting us.

She continued: "There are plenty of people who made money off us, lawyers living in fine homes. And everybody always wants to talk to us and talk about us. Where were they when we were fighting?"

"Yes ma'am," I responded. I said it over and over. I told her that I heard what she was saying, and, stopping myself from apologizing again, that I would pray for her and her community. I couldn't think of anything else I could say to her.

Apologies have three components, according to Samuel Oliner, a sociologist and Holocaust survivor. He calls them the three Rs: regret, responsibility, remedy. Regret involves recognizing the hurt and pain caused to another, an expression that such suffering should not have happened. Responsibility acknowledges the perpetrator's role in causing the harm. Remedy heals the damage and restores the hurt person to well-being.

Apology is a long-term commitment, an ongoing relationship, writes Aaron Lazare, who was a medical doctor and dean of the University of Massachusetts Medical School. He calls an apology a process that is often "arduous and uncomfortable."

Christine Bennett rejected my apology because it failed. On the phone that day in the Best Western, I offered only regret for what happened to her and her community, what Lazare calls an empathic "I'm sorry"—not an apology. I did not find a way to express my responsibility, my complicity with the harms she suffered, and I could not, as she pointed out, heal the damage.

An apology of that magnitude requires a lifetime commitment, a "way of being" kind of apology, as Yukiyo put it. My expression of regret to Christine Bennett only highlighted the holes where accountability and reparations should have been. A white supremacist, heteropatriarchal culture, built and organized around the notion that certain lives are disposable, is incapable of living out such an apology unless it transforms itself from the root.

: :

We drove into Mossville from the west on Old Spanish Trail, past Coach Williams Drive, named for LaSalle Williams, the beloved coach and then principal of Mossville High School. As far as I know he still lives in Mossville. We passed Mt. Zion Baptist

Church. The sign out front said, "Give God what is right, not what is left."

We crossed what used to be the town's main intersection with Prater Road, named for a Mossville family, and passed the Rigmaiden Recreation Center, named for another local family, with its community hall, swimming pool, and baseball diamonds. Behind it sat the former water works, its white tower rising up, MOSSVILLE written across it in the former school colors, maroon outlined in gold.

Among the cypress and pines sat a few houses on large lawns, some well kept, others abandoned with shuttered windows. But mostly what we saw were holes—empty lots with the house scraped off, only the concrete pad left.

Just past the Rigmaiden Center, we drove past the plants: SASOL on the left and Phillips 66 on the right. I held up my phone to video the tanks and towers, pipes and flares scrolling past. The hissing roar I had first heard in Freeport once again filled the air. I opened the windows to capture this sound, but the chemical, noxious stench overpowered us.

I first visited Mossville with Cobalt and Aster in summer 2019. I returned in the fall. I never located anyone from the community who wanted to talk to me, but I listened to the sixty or so oral histories of Mossville residents archived online, which were conducted by Louisiana State University and bankrolled by SASOL.

As I drove and wandered around, I had the words of people who lived here in my ears. I learned that in early days people hunted, grew food, and lived off the land. I learned that plumbing and electricity did not arrive until the mid-twentieth century, that the main road was not paved until the 1970s, that the town is named for James Moss, a former enslaved person who acquired the land through a federal grant in 1866. I learned that many in town had nicknames, that sometimes even lifelong friends didn't remember each other's given names, that the high school kids went to "record hops" in local canteens, small businesses that served soda and snacks. "We had sock hops, record hops, weenie roasts, and somebody was always throwing a party," Christine Bennett told her interviewer.

I heard stories similar to the ones told by Manning and Jessie about East End Freeport, a tight-knit community where people kept an eye out for each other. "Everybody took care of everybody's children," said Lenoria Braxton Ambrose. "Somebody could talk to you, whip you, send you home, and you might catch a whipping again."

I learned about the nightclubs, Joy Hill and Paradise Club, which brought in big-name acts like B.B. King, Ray Charles, and Etta James. I thought about Robert Hughes' work to recover the memory of the neighborhood Blues clubs in Houston, and the life that they nurtured. The names of the Mossville clubs offered promises that seemed too extravagant for reality—that within their walls could be found joy, could be found paradise. But I changed my mind listening to the oral histories.

For residents, this rural community tucked into the trees offered a haven from the surrounding racist terror. Black people came to Mossville from nearby communities for social and sports activities, where they could be safe from violence and the humiliations of segregation. No one locked their houses. "Some of my earliest childhood memories were there was nothing for us to be afraid of," said Ambrose.

: :

Delma and Christine Bennett did not sell their land in Mossville to SASOL. Christine lived there from the time she was born until 2010, when she was fifty-six. They moved to nearby Lake Charles because of Christine's health problems. She suffers a variety of ailments she said that doctors had not diagnosed. She still mourns the loss of her home, her neighbors. "I've been here now about six years, and I cry just about every other month...In Mossville...if I go to the store I'd stop off and visit somebody... We knew everybody and it was just a beautiful feeling...I said I wasn't going to cry, but nobody knows the pain," she told her interviewer, voice breaking.

For Christine, money does not heal. "You can't pay for our heritage and you can't pay for our health," she said. "How can you pay somebody for life?"

BLANKS

Both of my visits to Mossville started in the nearby town of Lake Charles at the Imperial Calcasieu Museum, a low, colonial-style brick building nearly inside the arms of a massive Southern live oak. Lake Charles and Mossville sit within the parish of Calcasieu, a French transliteration of an Atakapa phrase meaning "eagle cry." This region is the homeland of the Eagle band of the Atakapa-Ishak. European settler colonists decimated the Atakapans through murder, disease, and poverty, yet their presence permeates the place names and culture of southwest Louisiana and southeast Texas: in zydeco music, oyster pie, and the cured pork meat called tasso found in dishes like jambalaya. In 2006, Atakapan descendants gathered formally as a tribe for the first time in one hundred years. Tribal members are working to preserve their culture and language and to receive federal recognition.

Louisiana created Calcasieu Parish in 1840. It was the largest parish in the state, so people called it "Imperial," which means a span of earth and all who dwell there are subject to another's power. Before the U.S. acquired it, the land passed back and forth between the Spanish, French, and British empires, and from 1806 to 1821 it was designated "the neutral strip" between Spanish Texas and the United States, a refuge for many types of people looking to evade the laws of warring empires. The pirate Jean Lafitte used this area to smuggle enslaved Africans after the U.S. banned the importation of people as property in 1807.

On my second visit to the museum, a beautiful, warm fall day in November, I pulled into a nearly empty parking lot. I walked to the back of the brick, colonial style building to see the oak, its massive limbs curving all the way to the ground at points. This tree stood when the Atakapa-Ishak lived off the region's marshlands and forests. The museum claims it is three hundred seventy-five years old. It is named the Sallier Oak for Charles Sallier, a friend of Lafitte and an early European settler colonist, for whom the city is also named. He seems to have had little to distinguish him except for his whiteness. The city website describes him as "the first white man to build a house within what are now the city limits of Lake Charles" (in 1803), and the father of "the first white child born in Southwest Louisiana." The museum sits at the site of his winter house.

The Imperial Calcasieu Museum is a monument to the wealthy white timber families of the late nineteenth and early twentieth centuries. The front case contains a small collection of arrowheads and some outdated text about the Atakapa-Ishak, taken from a history published in 1952 by the Smithsonian called *The Indian Tribes of North America* by John R. Swanton. He writes of the Atakapa-Ishak in the past tense: "Although in 1907 and 1908 I found a few Indians who knew something of the old tongue, it is today practically extinct." This silencing, this erasure, belies the 1,800 members of six bands of the Atakapa-Ishak Nation living today who still must fight against a dominant culture that wants to blank them out. As the Nation's website asserts: "The Atakapa-Ishak are not extinct."

I could not find a word in the museum about Black people or any reference to plantations or slavery. I asked the museum director, Devin, about this. He was a young white man with lots of dark hair and a large moustache. He told me that slavery was not common in this part of Louisiana because it depended on the timber industry, which did not use slave labor. "Most slaves here were domestics," he told me, "they lived in houses, like family."

This erasure stunned me into silence. It may have been true that large slaveholdings were not common in the city, but that ignored the plantations around Mossville, seven miles to the west. Also missing was any evidence of the Mossville exhibit sponsored by SASOL. Devin informed me that the museum housed the exhibit in a separate annex and that I could not go there until the other visitors left, because he had to lock the main building and go with me.

The museum's other visitors, two middle aged white couples, asked if there were more to see than the one main room. Devin pointed them to the "war hallway" in the back, featuring the town's experiences of the two world wars. He did not mention the annex. I followed the couples back to the corridor leading to the bathroom. I was surprised to see a display dedicated to Claire Chennault, the man who tried to warn the U.S. and Britain about the Zero airplane during World War II.

Chennault was born on a farm on the opposite side of the state, but the local Air Force base, closed in the 1960s, was named for him, so the region claims him. The display effuses praise for Chennault as a "Louisiana farm boy" turned World War II flying ace, and a "great patriot." It includes one of his uniform jackets, and a painting of Chennault with a steely gaze and creased forehead. I also remember, though I did not take a photo of this, that the display included one of his wife's dresses, a beaded frock with an impossibly small waist.

I learn that Chennault divorced his first wife, Nell, the mother of eight of his children, in 1946, and wed Chen Xiangmei in 1947, when she was twenty-one and he was fifty-six. He met Chen, who became known as Anna Chennault, when she interviewed him in 1944 while working as a war correspondent in Shanghai. She spent much of the war struggling to survive as a refugee in one of the free areas of China, caring for her younger siblings after her mother died and managing to earn a university degree in journalism. Her older sister served as a nurse in Chennault's Flying Tigers squadron of U.S. pilots fighting in China, which is how she met him.

I wondered about Nell, the wife left behind. I found a feature about her in *Life* magazine from 1943 called, "Life Goes Calling on Mrs. Chennault: General's wife spends a busy Sunday at home in Water Proof, La." The piece features about a dozen photos:

Nell at church, teaching Sunday school, feeding chickens, serving a goose for dinner, and presiding over her children and grandchildren as they play and frolic on the farm. In only one picture, when she is holding a hymnal in church and singing, does Nell look directly at the camera. She is a stout woman in a dark dress and hat wearing thick glasses; she looks slightly startled at the lens looking back at her.

The article reports that seven Chennault children, six boys and one girl, had already grown and left home, most for military service. Only fourteen-year-old Rosemary remained with her mother and the family's longtime "cook," a Black woman named Annie Dorty whose portrait is included in the spread; she holds the general's grandson and namesake on her lap.

We learn the farm has forty-nine acres and features a poultry flock of fifty, two cows, a horse, an orchard, and gardens. What we know of Nell comes through others; she has no voice. The anonymous and somewhat condescending reporter calls her "a distinguished person in her own right" and claims that "the people of Water Proof, La. (pop. 592)...are as proud of her as they are of the general." It praises her as "a faithful worker in the little old Water Proof Methodist Church" and notes "the quiet charm and friendliness for which the community loves her."

We learn that her son Pat and his wife (parents of Claire Lee Chennault, II) "believe, with the community, that 'Mrs. Chennault sets as good a table as you'll find in this part of Louisiana.'" Annie Dorty gets no credit, although she is described as "cook" and Nell must have needed help. We learn that in addition to maintaining the farm and household and keeping up correspondence with her "menfolk" off at war, Nell worked for the Red Cross and was "chairman" of the community's war-bond drive.

The general's divorce from Nell dropped her from history. His new wife became more famous. After the war, Chennault and Anna lived part-time in Taiwan and part-time in Monroe, in central Louisiana, in a neighborhood that did not allow non-whites, in a state that forbade miscegenation. Chennault's celebrity as a war hero buffered them from this racism. After he died in 1958 of lung cancer, Anna moved to Washington, D.C., where she built a career as a political power broker and founder of an air cargo company called The Flying Tiger Line. She also fought against communism by supporting conservative, anti-communist Republicans.

In 1968, Anna served as a secret channel between presidential candidate Richard Nixon and South Vietnamese officials, helping to derail peace talks advocated by President Lyndon Johnson. Nixon feared a peace brokered by the Democrats might tip the election in their favor. In return, Anna expected a prominent role in Nixon's administration. Behind her back, Nixon called her "Dragon Lady." He denied her a position because he was afraid she would reveal their collusion. Her 2018 obituary in *The Washington Post* calls her "a figure of glamour and mystery."

I remember none of this from the exhibit about Chennault. It may have been there, and I erased it myself, intent on getting to what I had already decided was the heart of this investigation, which is how I often end up blanking out the lives of women.

SILENCE

The first object that greets visitors to the Mossville exhibit at the Imperial Calcasieu Museum is the white marble tombstone of Thomas Rigmaiden, slaveholder. It reads: "Native of England, Born October 29, 1786, died July 2, 1865." Like the word "plantation" on the plaque in Lake Jackson, this white slab inserts itself like a tooth, gnawing its way from the past, through the present, to the future. It is death, its chewing whiteness, the relentless racist violence that the people commemorated in this museum worked for generations to escape.

Above the death tooth, the left wall lists the founding families of Mossville. Some took the names of slaveholders, or they inherited names from slaveholder parents, or both: Moss, Perkins, Rigmaiden, Towner, Williams, Lyons, LeDoux, Vincent, Braxton. Below each family name are portraits and family photos: Lenoria Braxton Ambrose as "Miss Mossville 1964" in a white gown with a cape, scepter, and tiara. Her grandmother Loni Braxton Perkins sits unsmiling in a black suit and hat with white gloves, handkerchief, and pearls. Dave and Lula Lyons in wedding attire—they ran one of the canteens for teenagers.

The annex building is a smallish square, maybe twenty by twenty feet. The exhibit is somewhat sparse. The back wall contains items from the Mossville school building, which SASOL purchased and then demolished. It includes stones from the school's façade, a long green chalkboard, an American flag, intercom, and school bell, and, behind glass cases, a gold and maroon band uniform, cheerleader outfit, and letter jacket. A maroon and gold rug features the school mascot: a pirate with a skull and crossbones on his hat and a sword in his teeth. A yellow sign lists "RULES FOR OUR LUNCHROOM" in black marker, including "1. No Gum Chewing" and "2. No Loud Talking" and, across the bottom, "Let's Make Our Lunchroom One Of The Best!"

In the oral histories, many residents remembered the school as the pride of their community. The town built the school with grades one through twelve in 1955. People remembered teachers who pushed them hard to excel, to be better than their white peers in surrounding towns, even though they used tattered, secondhand textbooks. "Our books were not usable...but we used them," said Evelyn Gasaway Shelton. They remembered ranking among the best in the state for academics. Coach Williams took Mossville High to state championships in basketball, track, and football. A few Mossville residents went on to be professional athletes in the sixties.

The exhibit includes a few yearbooks from these years. The 1961 photos list student nicknames, like "Bush," "Glo," and "Tallboy." Each student also includes an ambition: engineer, vocalist, seamstress, nurse. High school clubs included chorus, math, Home Ec, biology club, and library club. The 1956 yearbook features a list of "Prophecies," with the names of students who would fulfill them, including: "Who will star in Hollywood's largest picture? Audrey Prater; Who will take the place of Albert Einstein? Edgar Smith; Who will sing at Metropolitan Opera House? Tressie Valentine." The section ends: "May you walk confidently into the future and fulfill your destinies."

On the right wall of the Mossville exhibit is a computer station with headphones to listen to the oral histories. Aside from these testimonies, stored away in digital files, the voices of residents are almost entirely absent from the exhibit.

But silences and blanks also permeate the oral histories. Residents in general seem reticent to talk about negative things for the record. Most of the people conducting the interviews (some seem to be employees and some students of LSU) ask a set of preset questions about family life, religion, home remedies, gardening. Few stray from the script or delve into controversial topics.

Mossville High graduated its last class in 1969, when the move to desegregate closed the high school and students had to bus to nearby Sulphur and Westlake. Brenda Cole Jones was among the first to integrate Westlake High School. In her oral history, she remembered arriving on the bus to find white parents out front with nooses, though she doesn't use this word. "There were parents that were outside. They had ropes with...And they were shaking them...and saying, 'This is for your kind of people.'"

Christine Bennett lamented that she missed out on her high school education entirely because integration was so violent, with kids fighting each other every day.

"We begged them," she said. "Why couldn't Westlake come to our school? Then after we got to Westlake High and...Do I say just really what happened? You want me to say that?"

This silence, its functions are multiple. Imposed on residents from the outside—by SASOL, by the institution of the museum and its interpretative language, by the confines of the oral history project—silence keeps the narrative of Mossville residents within acceptable boundaries, minimizing the threat to status quo power structures. Residents also choose silence, like Christine's refusal to talk to me, as a way of keeping themselves safe. This made me think of the storyteller Esther Stutzman of the indigenous Kalapuya of the Willamette Valley in Oregon. How she does not tell certain of her people's stories, guarding them as a form of wealth that an oppressive culture cannot take.

Daren Dotson is one of the few people to talk openly about the plants, in response to a stock question near the end of the interview: "What's the most important thing you want people to remember about Mossville?" Daren did not live in town but came in summers from Maryland. He says, "Well, it's a known fact around here that the people of Mossville been fighting these plants for years, and years, and years...it's a known fact that a lot of people around here died of cancer." Asked by the interviewer why people seem hesitant to talk about it, Daren says, "It's kind of like throwing water in the wind. It's going to blow back on you."

Throughout their oral history, the Bennetts speak openly about struggles with pollution. Delma remembers the smells. Each plant had a specific odor—rotten eggs, ammonia—and when it got bad enough the school would move the kids in from the playground. Christine notes that when integration started, they had to drive through the plants to get to Westlake High, along the same road we had taken past SASOL. They drove with the bus windows closed because of the pollution. "We had to deal with this every day to go to high school," she said, "and the kids are still passing through those industries to get to school every day."

Jawana Mayo Huntsberry mentions pollution at the end of a two-hour interview when the interviewer specifically asks her about it. She describes regular explosions and remembers once having to shelter indoors as flecks of something fell from the sky. Brenda Cole Jones, who spent her career working at one of the plants, also remembered explosions shaking her bed at night. "We didn't have streetlights but when the flares would go real bright we could run outside and play a little longer," she said. "We never thought it...could have made us sick."

The oral historians did not speak to Debra Sullivan Ramirez, long an activist leader in the community, but she is featured in a 2014 book called *Women Environmental Pioneers of Louisiana.* She remembers sitting on her grandmother's porch swing and being lulled to sleep by the humming sound of the chemical plants. She also remembers the police going house to house in the middle of the night telling people to evacuate because of explosions and chemical releases. "We got to the point where we were sleeping in our clothes," she said. "Little girls had nightgowns sometimes that were thin, and when they told us to leave, we didn't even have time to grab a housecoat...We had to go to school the next morning, sometimes we would be very tired."

Dorothy Hartman Felix can trace back seven generations of her family, most from this part of Louisiana. She was in her late seventies when she finally moved to the Houston area to be with her children and grandchildren, one of the last residents to accept the buyout. I tried to speak with her several times, but she always politely put me off. Felix served as president of the Mossville Environmental Action Network. She traveled to Washington, D.C., to testify about the pollution problems in Mossville and played a leading role ten years ago in winning the first-ever hearing granted to U.S. citizens by the Inter-American Commission on Human Rights. Mossville residents and activists petitioned the organization to investigate the U.S. government's violations of their civil rights because of its failure to address pollution. Felix's oral history interviewer does not ask about this.

Felix is interviewed along with her childhood friend Evelyn Gasaway Shelton, who mentions this history in a coded way at the end of the interview, describing notable Mossville citizens, including "Dorothy Hartman," a "well-traveled community activist." Felix brings it up herself at the end of the interview, in response to the stock question regarding what they want people to know about Mossville. She replies, "There is so much that can be said and that needs to be said and should be said." She notes that the petition is one of the main things she wants people to know about. "I have it right here!" she says. "I brought it so you could see I'm telling you the truth [everyone laughs]...We physically went over [to Washington, D.C.] and handed those documents to that organization."

Her interviewer responds, "Alright, well we're going to go ahead and close this up..." The last words recorded are Felix being cut off, saying again "I have the..."

ALIVE

I pulled into the parking lot of the Heaven on Earth BBQ & Seafood Restaurant at about 1 p.m. on a Saturday. The restaurant is on Prater Road, the only commercial establishment left in Mossville. It is inside a square brick building that looks like a house, with a front counter and about six tables. I was the only patron. A K-pop video played on the TV. I ordered brisket, which came with potato salad, dirty rice, baked beans, and a slice of white bread. I bought a hot pink t-shirt with the restaurant's logo: a grinning pig with wings wearing an old-fashioned leather aviator's hat and a blue scarf and holding up barbeque tines speared through a rack of ribs.

Madison worked the register. She looked to be in her twenties. She told me she was the goddaughter of the owner, a Rigmaiden. "Oh, I know that name, it's famous here," I said, but I did not mention writing a book or anything else about me. I asked about business. She said it's been very good, busy with people from the plants. She told me about the oral histories. "They didn't even talk to everyone," she said. "But maybe some people didn't want to talk because, you know, it hurts."

Maybe this was why I also didn't speak about myself. I had come to this town to find things out, to learn things, but I wanted to honor the boundaries, the silences that people chose for themselves, the divisions I could never cross.

I stashed my Styrofoam takeout container in the SUV and took off walking up Prater Road toward Old Spanish Trail. The overgrown sidewalk passed the concrete pads of former houses and driveways, and it appeared some people might still have been living back behind the curtain of trees. A tanker truck roared past, the driver slowing for the stop sign at the intersection and staring at me, craning his head back to get a better look.

I felt on edge in this place, like I was being watched. I took video as I walked, holding my phone surreptitiously at my waist. The bright sun, the *chock, chock, chock* of my footsteps, the sound of cicadas, passing trucks.

I crossed Prater Road and walked along Old Spanish Trail: more house foundations, a concrete front walk with the remains of blue-painted brick porch supports, a basketball hoop along an old driveway, with TEAM USA written in faded blue script across the backboard. I crossed Old Spanish Trail and walked toward the Rigmaiden Recreation Center. I looked through the fence at the baseball diamonds, videoed a monarch butterfly flittering through the chain link in front of me. I wandered to the front by the swimming pool. A sign read: "NO PROFANITY ALLOWED ON PROPERTY." The place seemed deserted, though as far I know the recreation center is still in operation.

I crossed the street again and walked further, toward the new overpass and the entrance to the plants. Just before it, a dirt road led off to the right toward the Morning Star Cemetery. I had visited the cemetery in summer with Cobalt and Aster, when it was more than 100 degrees. I videoed that trip also, the air thick with cicadas and heavy with humidity. We wandered, sweating, among the graves, all of them enclosed in concrete containers above the marshy ground. The swamp—water, cypress knees, thick moss-hung trees—lay just beyond the chain link fence. I felt its presence, a watery and ancient breathing.

The cemetery used to belong to the Morning Star Christ Sanctified Methodist Church, where Dorothy Felix and Della Dotson worshipped, in a church Della said her father Josh Rigmaiden built. Hurricane Rita destroyed it in 2005, and the community lacked the money to build it back. The cemetery has graves dating from the nineteenth century to the present; it is still an active burial place. Della Dotson's mother used to maintain it. She remembered her mother recruiting people to help her paint and repair the graves and mow the grass. "Every two weeks my mom took her own money and paid somebody...to cut the graveyard," Della remembered. Now SASOL claims to maintain it, but I wonder how well. On my second visit, a tree had fallen and crushed a section of fence.

Many familiar names fill the cemetery, including Tom Hartman, Dorothy Felix's grandfather, who lived from 1849 to 1922. There is a stone for David Dotson, Jr. and Darlandus Dotson, who lived for one day, from September 19 to September 20, 1969. Many Rigmaidens also are buried here, including the town patriarch Joshua (J.A.) Rigmaiden, who lived from March 8, 1876 to November 19, 1957.

Standing in that cemetery, surrounded by the remains of individuals who built this community, who survived here, thrived, despite the brutal and relentless racist violence assaulting them across generations, I felt the weight of what Christine Bennett meant when she said she wanted a memorial, something left for her community.

She wanted just compensation and healthcare, and she wanted broader reparations, including some physical testimony: not a museum exhibit in a nearby city, not digital ones and zeroes, but here, a testament to the lives lived on this soil, the life of this community, and what it meant.

How it hurt.

How it continues to.

Remember

: :

Mossville is not unique. Since the 1980s, industrial expansion has erased at least five other communities founded by formerly enslaved Black people in Louisiana: Diamond, Morrisonville,

Revilletown, Sunrise, and Wallace. Often what is left of these communities are their cemeteries, and these are also at risk. Residents of the majority Black St. James Parish had to win a court order to hold a Juneteenth prayer at the graves of their ancestors on land now owned by Formosa.

Robin McDowell, a scholar and community organizer, wrote an article for the *Boston Review* in 2019 about the struggles of people from the former Revilletown to maintain access to their cemetery. A corporation that makes polyethylene, vinyl, and other products, until recently known as Axiall, bought out Revilletown in the early 1990s, and now claims it also owns the graveyard. Residents can only access their ancestors' graves, or bury their dead, with company permission.

Recently Westlake, a company with more than fifty plants across the U.S., Europe, and Asia, bought Axiall. Its largest vinyl plant is in Sulphur, Louisiana, seven miles south of Mossville.

Why, asks McDowell, would a corporation with global reach and influence go to court to assert its ownership over "a half-acre parcel of grass and bones"? Her answer: Profit-generating machines, corporations like Westlake, Dow, and SASOL, depend on an "economic, social, cultural, legal, and racial regime that spreads over land and across time." Maintaining this profit-generating order, protecting the status quo, means "nothing can be allowed to escape, not even a parcel of grass, because then anything could."

Grass comes from an ancient root that means "to grow, become green." It is synonymous with life, which always, across time, escapes the bonds that would contain it, like the ancient surrounding swamp. It receives water rushing across the land and slows it, draws it down deep into sediment and root, letting it eddy and pool, the growth here pulling out not only poisons and pesticides but also nutrients, often brought from hundreds of miles away, from the region called the U.S. heartland, to feed its green life, the life that luxuriates everywhere around Mossville.

what has been dead, writes the poet Audre Lorde, *is now alive*

TRUST

On November 1, 1952, the U.S. military exploded the world's first thermonuclear device, codenamed Mike. It was made possible in part by the new plastic, polyethylene, the same plastic that makes up my car part, and now the most common plastic on the planet.

The military ignited Mike at Enewetak Atoll in the Marshall Islands. The one hundred fifty-seven inhabitants of the atoll were floating in the ocean, crammed onto a U.S. Navy boat.

The U.S. government had actually moved the ri-Enewetak five years earlier in order to test nuclear weapons on their home islands. The military chose Enewetak Atoll and Bikini Atoll, two hundred miles east, because of their distance from the U.S. and other large population centers. In their eyes, the islands were remote and unimportant: blank specks of land in the ocean's shining blank.

The U.S began its nuclear testing program on Bikini in 1946 after evicting its residents. The military used the first test series, codenamed Crossroads, to demonstrate to the world the power of nuclear weapons. The government issued dozens of press releases and brought hundreds of reporters on boats to film and photograph. People around the world heard the tests broadcast live by radio.

The military set off the second explosion in this series, codenamed Baker, underwater in Bikini Lagoon to see how it would affect the ships anchored there. The underwater explosion kept the radiation from dispersing into the atmosphere, leaving the lagoon dangerously contaminated. The head of radiological safety for the tests, Dr. Stafford L. Warren, warned the Navy that this would happen, but the military went ahead anyway.

The authors of *Bombing the Marshall Islands: A Cold War Tragedy* write, "A few millionths of a gram of radium ingested by a human being can be lethal. One hour after the Baker shot, the amount of radioactivity in Bikini Lagoon was equal to 5,000 *tons* of radium... it has been aptly described as 'the world's first nuclear disaster.'"

The U.S. had free range to do what it wanted in the Marshall Islands. After World War II, the United Nations parceled out former colonies in the Central Pacific and Africa to victorious nations as "trust territories." The U.S. governed the Trust Territory of the Pacific Islands, which included the Marshalls.

The U.S. government did not want the UN to limit its activities in the Pacific, so it secured a special status for the region as the world's only "strategic trust," overseen not by the entire UN General Assembly, as with all the other trust territories, but by the smaller UN Security Council, where the U.S. had veto power.

Territory means just a spot on Earth. Trust in this sense is legal: something one does for another, to care for a thing for the benefit of someone else.

Care means to feel concern, to trouble oneself, to grieve with or for another, from a root that meant "to cry out."

The U.S. did not entangle itself this way with the Marshalls. Its interest was military, as a "proving ground" for nuclear weapons. The U.S. governed the region under military security throughout the 1950s. No one could enter without permission; no non-U.S. citizens could enter at all. The nation used the arrangement to try and control information about its nuclear tests and block access to Russia and China, the rivals it feared most.

This left the Marshallese living in their homeland inside a security state that largely erased them; they were not the threat. The Marshall Islanders were obstacles to be moved out of the way, a potential public relations problem for the U.S. Atomic Energy Commission (AEC), which oversaw the nuclear tests. As the authors of *Bombing the Marshall Islands* write, "It was not that [AEC head Lewis] Strauss looked down on the Marshallese; he just looked through them."

: :

In the twelve years from 1946 to 1958, the U.S. Government conducted forty-three nuclear tests on Enewetak Atoll and twenty-three on Bikini Atoll. By the time the U.S. followed Russia's lead and ended atmospheric testing in 1958, it had exploded 80 percent of its atmospheric nuclear tests on these two atolls. Nearly half the tests took place during a single year, 1958, as the military rushed to complete tests before the moratorium. The U.S. exploded all of its largest thermonuclear weapons in the Marshalls, reserving smaller tests for the desert of Nevada.

In 1980, after thirty-three years in exile, the ri-Enewetak resettled the southern portion of their atoll—the least contaminated part, on the islands of Enewetak, Medren, and Japtan—less than half their original land area. The U.S. government built homes, community centers, churches, and other facilities and planted thousands of coconut, breadfruit, and pandanus trees. Scientists concluded that even on these islands, the soil would need to be treated and monitored for radiation for the next century. Because of the dangers, the ri-Enewetak still must import their food.

As for Bikini Atoll, some residents returned in the early 1970s. The Atomic Energy Commission claimed the land contained "virtually no radiation." But in subsequent years, the U.S. government began to find contamination in local foods and in the bodies of the people. Concerned by the conflicting reports, the Bikinians sued the U.S. government, demanding a complete scientific study of the safety of the atoll.

Meanwhile, regular surveys found radiation in well-water that exceeded maximum limits, and an eleven-fold increase in radioactive cesium in the blood of Bikinians. Calling the increase in cesium levels "incredible," the U.S. Department of Interior evacuated the Bikinians in 1978.

Today Bikini Atoll contains some of the planet's highest radiation levels. It remains uninhabited, except for a half dozen caretakers who monitor radiation in six-month shifts to limit their exposures. Bikinians live in diaspora, scattered across the Marshall Islands, the U.S., and the world.

HOLE

After the Baker test in 1946 left Bikini Lagoon too dangerous for military personnel, the U.S. shifted its testing to Enewetak. The ri-Enewetak had already suffered years of warfare and displacement. Japan took control of the Marshall Islands after World War I. In the 1930s, the Japanese government established bases there, planning to use the islands as stepping-stones to enlarge its empire and attack the U.S. The military conscripted Marshallese people to build airfields, airplane hangers, and docks.

As the war dragged on, some Japanese treated Marshall Islanders with brutality, forcing them to harvest food for them, and killing starving Marshallese who kept food for themselves. They also executed English-speaking Marshall Islanders and Christian missionaries for fear they were spies for the U.S.

After the battle of Midway, when the tide of war turned, the U.S. military began attacking the Marshall Islands to push the Japanese out. On February 17, 1944, the U.S. invaded Enewetak. The military had already bombed the Marshalls relentlessly, destroying every Japanese airplane on the islands. U.S. troops came ashore by the thousands and fought in brutal hand-to-hand combat with Japanese soldiers.

The Japanese allowed some civilians to leave before the invasion, but those unable to leave suffered. One ri-Enewetak described it:

> If you wanted to go, you had to dig, for there was no other place open for escape. From dawn to dusk they shot at this islet. At the finish, there was not a single coconut [tree] standing...All of us were in the holes. Anything not in the holes disappeared.

The survivor describes U.S. soldiers arriving, looking inside the hole, and throwing in a hand grenade. "Then it burst; the whole shelter was torn apart. So powerful was the thing one could never stand. Earth fragments struck us, but others in the other half [of the shelter], they died."

The word in Marshallese for death is *mej*. From the Marshallese-English Dictionary:

> *mej_1*: dead; numb; death; disease; illness; peril; plague; wrath.
>
> 1. *Ekeiñtaanan mejin (mijen) lōḷḷap eo*. The old lady's death was torturous.
> 2. *Emmejmej nājin lieṇ*. Her children all die.
> 3. *Iar lo mijen inne mokta jān an mej*. I saw his spirit yesterday before he died.
> 4. *Mejet ṇe aṃ?* What's your illness?

: :

Recovery for those who survived the war had just begun when the U.S. Navy arrived, four days before Christmas in 1947, and moved the ri-Enewetak from their homeland to a nearby atoll named Ujelang. As military personnel ushered them onto the boat, one translator told them, "You cannot protest or fight. You are like a rabbit fish wriggling on the end of a spear. You can struggle all you want, but there is nothing you can do to escape."

In the days before the Mike test, nine thousand U.S. military personnel and two thousand civilians swarmed Enewetak, preparing to ignite the fire of the sun on Earth. No one knew for sure what might happen. To be safe, the U.S. Atomic Energy Commission gathered the ri-Enewetak from Ujelang onto a boat and moved them further from the blast site. The islanders watched the explosion on their homeland from the deck of the ship. The U.S. forbid them from describing it.

Others did describe what they saw; thousands of service members, technicians, and scientists also watched from ships thirty miles from the blast site. News began to leak out despite official silence. A few years later, the government released a "sanitized" film of the explosion for TV, with sections that the military deemed classified removed.

People around the world watched Reed Hadley, narrator of the TV crime drama "Racquet Squad," wander around the Mike control ship *Estes* in a sweaty khaki shirt interviewing military officials and scientists in the lead up to the blast. Hadley lights a pipe and drinks from a white porcelain teacup as members of the testing task force give canned responses to his questions.

Foreboding orchestral music announces the detonation. The edited sequence shows a burst of light, lightning streaking upward from the ocean, and the roiling fireball expanding. The shockwave ripples across the surface of the water to the ship as the music crescendos, and the mushroom cloud spreads across the sky.

Even with the theatrics, the film doesn't capture the scale and magnitude of that blast, three times larger than all the explosives used in World War II put together. In his book *Dark Sun*, the historian Richard Rhodes compiles eyewitness reports and scientific data about the event:

> [The blast] expanded in seconds to a blinding white fireball more than three miles across...the crews of the task force, thirty miles away, felt a swell of heat as if someone had opened a hot oven, heat that persisted long enough to seem menacing...
>
> Swirling and boiling, glowing purplish with gamma-ionized light, the expanding fireball began to rise, becoming a burning mushroom cloud balanced on a wide, dirty stem with a curtain of water around its base

> that slowly fell back into the sea. The wings of the B-36 orbiting fifteen miles from ground zero at forty thousand feet heated ninety-three degrees almost instantly...
>
> At its farthest extent, the Mike cloud billowed out above a thirty-mile stem to form a huge canopy more than one hundred miles wide that loomed over the atoll.

In place of the island Elugelab, the bomb left a crater more than a mile wide and two hundred feet deep filled with seawater, "a dark blue hole punched into the paler blue of the shallow lagoon."

: :

The word hole is related to hull, hell, and hold. It comes from a root that meant "to cover over, conceal, save."

The Urban Dictionary defines slang uses for hole: solitary confinement; any wretched and unpleasant place; mouth; anus; vagina, and by extension, female.

It seems Ulam did not attend the Mike test that, as he noted, "more than confirmed the possibilities" of his idea. He wrote, "After completing this theoretical work [on the bomb], I considered my job done."

He recorded some thoughts on the morality of his effort:

> Contrary to those people who were violently against the bomb on political, moral, or sociological grounds, I never had any questions about doing purely theoretical work. I did not feel it was immoral to try to calculate physical phenomena. Whether it was worthwhile strategically was an entirely different aspect of the problem—in fact the crux of a historical, political, or sociological question of the gravest kind.

He apparently considered that question outside his purview as a mathematician. He wrote that such issues "had little to do with the physical or technological problem itself." This is, of course, an erasure, a denial that Ulam's inquiry took place within these "historical, political, and sociological" structures, and cannot be separated from them. They are entangled.

Edward Teller did not attend the test either. Frustrated and resentful about what he considered attempts at Los Alamos to block his progress, he convinced the Atomic Energy Commission to let him start his own laboratory in California (now Lawrence Livermore National Lab, still one of the nation's three primary nuclear weapons labs).

The day of the test, Teller—who would become known to history as the "Father of the H-bomb"—went to the basement of the geology building at the University of California, Berkeley. He sat in a darkened room staring at a light point on a seismograph. At the appointed time, the shock wave from Mike rippled five thousand miles through the surface of Earth and caused the light point on the seismograph to jump. Teller telegrammed three words to the anxiously waiting scientists at Los Alamos: "It's a boy."

CIRCLE

Iep Jãltok: Poems from a Marshallese Daughter is the title of the first book by Marshallese poet Kathy Jetñil-Kijiner. An epigraph gives the definition of the Marshallese phrase:

> *"A basket whose opening is facing the speaker." Said of female children. She represents a basket whose contents are made available to her relatives. Also refers to matrilineal society of the Marshallese.*

The first piece in the book, "Basket," is a visual poem defined by the hole at its center. It echoes the shape of baskets woven by Marshallese women and also the reef-encircled lagoons of Kathy's home islands. The surrounding text can be read in multiple orders: down one side and up the other, in a circle; down one side, then down the other, meeting in the middle; or by reading across the gap. It merges images of woven baskets with female bodies in ambiguous, multiple renderings: woman-basket is brimming with offerings, connected to land and ancestors; and it is scraped bare, a receptacle for litter. The poem contains a smile, both as a shape within the visual poem and in the words of the text. As a shape, it mirrors itself, forming the rim of the basket. As text, it remains as resonant and indeterminate as the blank it borders. Some of the possible readings:

> *i fell asleep / dreamt / / my smile / was merely / a rim / woven / into my / / face*
>
> *face / / into my / woven / a rim / was merely / my smile*
>
> *my smile / was merely / a rim / waiting / / to be / woven / into my / woven / / face*

Kathy is the daughter of former Marshall Islands President Hilda Heine, the first female president of the islands. Hilda is the granddaughter of Carl Heine, an English missionary executed by the Japanese military on April 20, 1943. I came across a blog called Marshall Island Voices that includes an anonymous post from almost exactly seventy years after Carl Heine was murdered. It is by someone who identifies him as a great-grandfather. The person writes:

The Japanese...were trying to prevent missionaries from preaching the word of God around the Marshall Islands. They warned people that whoever was preaching...they would have them beheaded or hanged. The Japanese soldiers found out that there was an old man who persisted in preaching even though he was aware of the consequences. Before they arrested him, he put all his children, grandchildren, and wife in a hole where they could hide from the Japanese troops so they would not be killed.

Kathy is an acquaintance of mine through poetry and through Yukiyo, who is close friends with Kathy. She lived in Portland from 2016 to 2019 before returning home to the Marshall Islands with her five-year-old daughter. I wrote her to ask about who might have posted this note, but she did not respond. She was in the midst of campaigning for her mom's re-election. Kathy posted a photo to Facebook recently with three generations of Marshallese women: Kathy's mother, the first person with a Ph.D. in the Marshall Islands; Kathy, the first Marshallese published poet; Kathy's daughter, the subject of a poem that Kathy performed before a global audience in 2014 at the UN Climate Summit. I think about the shelter of that hole. The continued generations it made possible.

Circle as zero, as hole, a resonating potential—

It is the ouroboros snake biting its tail, whirling, as in the Houston artist Julia Barbosa Landois's work, through a deep blue sky. It encircles the world, boundary between chaos and order, planet holder.

: :

The earliest known depiction of an ouroboros comes from the tomb of King Tutankhamun in Egypt, three thousand years ago. The symbol stood for the cyclical nature of time, its eternal return, represented by the rising and setting sun. The past seems to disappear, but it only changes form, becomes "re-membered" inside the body of the snake. The ouroboros is one of the oldest mystical symbols. It spans time and cultures: ancient India, China, South America, Scandinavia.

The idea that time is linear, moving forward relentlessly, is European, underscored by Newtonian physics and imposed on others by colonial structures. This concept of time does not exist in the world-making of cultures that, like the ancient Egyptians, interpret time as circular and all times as present, as with the old and new sun in its constant cycles. The past is the seed we living beings carry in our bodies. So is the future.

How shall we remember you?

The last piece in Kathy's book is also titled "Basket." The book forms an ouroboros, circling back on itself. The past bursts forth in these poems, in her grandmother's tongue cancer, in her niece's death from leukemia. "Basket" fuses the indelible past-present of violence and suffering—"a seabed / to scrape / a receptacle to dump / with scraps"—with other potential past-futures—"a reef / of memory / / your womb / the sustainer." The resonating blank in the center holds in tension these possibles.

PULSE

The pulse of radiation from atomic bombs is global, but much of the fallout landed on those closest to the blasts. People absorbed it through their skin, breathed it in, or consumed it in water and food. Radiation concentrated in the flesh of coconuts and crabs, common Marshallese staples. The invisible energy radiating through the bodies of Marshall Islanders damaged DNA molecules and triggered a cascade of disease and suffering through generations: cancer, leukemia, degenerative conditions, growth retardation, and birth defects. People reported women giving birth to "'jelly babies,' unformed fetuses that were not recognizably human."

More Marshallese people now die from cancer than any other disease except Type 2 diabetes. This also has links to the testing. It comes in part from eating processed food imported from the U.S. after local food became too contaminated to consume. Marshall Islanders suffer an official list of thirty-seven radiation-related cancers and diseases, but thyroid is among the most common. The thyroid is a butterfly-shaped bit of tissue that enfolds the windpipe. It helps regulate nearly every physical system: digestion, circulation, cognition, muscles, bones, mood. The small organ concentrates iodine from the body, which it uses to activate and release hormones. Nuclear explosions produce a lot of radioactive iodine that the thyroid dutifully gathers.

: :

Kianna Juda Angelo invited me to her house in Lake Oswego, Oregon, just outside Portland, in February 2020. She was home for a few days between trips to the Marshall Islands and Denmark, where her husband's family lives. She greeted me with a hug and brought me up to the kitchen. Half-packed suitcases covered the dining table. She filled a platter with food—"You'll learn we

Marshall Islanders feed people!" she told me—and started a fire in the fireplace against the rainy Oregon chill.

Kianna was born in the Marshalls and adopted by a U.S. family. She grew up in the Pacific Northwest. Her adopted family kept information about her home islands from Kianna. Her mother burned letters to Kianna from her biological parents in the Marshalls.

As an adult, Kianna started a nonprofit called Living Islands to support Marshallese communities both in the Marshall Islands and the U.S., and to foster exchanges between the two cultures. The work connects her to her birthplace and her biological family. She has land in the Marshalls inherited through her biological mother in this matrilineal culture, and she beams when she speaks about someday building a place to live there. But the radiation pulse beats in her also.

"The whole time I was growing up, I went to maybe six funerals," she said. "Now, I get invited to six funerals a month in the Marshalls." She told me that one of her half-brothers died of thyroid cancer, as did a niece who was in her twenties. "I get invited to more funerals than I do birthday parties."

All of Kianna's four children, ranging in age from eight to twenty-five, have heart problems. "We get sick more easily than my husband and our illnesses last longer," she said. "I worry about my children, and I worry about my future grandchildren."

: :

I met with Kianna just before the COVID-19 pandemic took hold in the U.S. We had no idea what was in store. The Marshall Islands remained virus-free into the summer, thanks to early action by the Marshallese government, including a ban on all incoming travel, leaving the islands quarantined in the vast ocean.

In the U.S., however, Marshallese people were catching the disease and dying at high rates. In Oregon, Pacific Islanders and

Native Hawaiians had the highest per capita infection rate: ten times higher than the state average, and twenty times higher than whites.

Northwest Arkansas has the largest concentration of Marshallese in the U.S. outside of Hawai'i. Many work in the giant Tyson Berry Street poultry processing plant in Springdale. In this region, Marshallese people make up three percent of the population. By July 1, they accounted for half of the deaths from coronavirus.

"The community is devastated," Eldon Alik, Consul General for the Springdale consulate of the Republic of the Marshall Islands, told the *Arkansas Democrat-Gazette*.

Manumalo Ala'ilima laid out the risks and barriers that Marshallese and other Pacific Islanders face. They co-founded and serve as board chair for UTOPIA PDX, a group that advocates on behalf of queer and trans Pacific Islanders. In an interview with Oregon Public Radio, Ala'ilima noted that information about the virus was not reaching communities because it's not in the right languages or not culturally appropriate.

People from Marshallese communities often work in so-called "frontline" jobs, the jobs that keep other people's bodies and the rest of the economy running: meat processing, warehouse work, healthcare, security. Compounding this, Pacific Island cultures are communal, and many live in multigenerational, multifamily homes, increasing the risk of infection. "It's not just your nuclear family, according to Western standards," said Ala'ilima.

Underlying all of this are the health problems already plaguing the community, and their lack of access to healthcare. "COVID-19 is shedding a light on something we've always known existed," said Ala'ilima. "The health disparities with regard to heart disease, diabetes, cancer...your immune system...is already threatened."

Under a 1986 treaty called the Compact of Free Association (COFA), residents of the Marshall Islands, Palau, and Micronesia can legally work in the United States. The agreement included access to health benefits through Medicaid. Ten years later, President Bill Clinton signed the so-called "Personal Responsibility and Work Opportunity Act," also known as "welfare to work," which built on years of anti-welfare propaganda and stripped many of those living in poverty of benefits, including people from these nations. They can legally work in the U.S., and pay taxes, but they can't receive federal benefits.

Marshallese generally avoid doctor visits until they are severely ill, said Dr. Sheldon Riklon, who works at the University of Arkansas for Medical Sciences in Fayetteville and was raised in the Marshall Islands. "They are already worried about their bills."

"A lot of families out here are seeing multiple deaths within the family. It's been heartbreaking," Kelani Silk told the *Orange County Register* in Southern California, another place where Marshallese suffered greatly from COVID. Silk lost her uncle and a cousin to the virus.

The deaths of elders are particularly devastating for a community in diaspora. "These are individuals who hold cultural wisdom and information that many of us who are born and raised in this country don't have yet," said 'Alisi Tulua, program manager for the Orange County Asian and Pacific Islander Alliance and part of the national Pacific Islander COVID-19 Response Team. "If we lose our elders, our stories, traditions, and connection to our roots may be lost with them."

1. *Ekeiñtaanan mejin (mijen) lōḷḷap eo.* The old lady's death was torturous.

2. *Emmejmej nājin lieṇ.* Her children all die.

3. *Iar lo mijen inne mokta jān an mej.* I saw his spirit yesterday before he died.

4. *Mejet ṇe aṃ?* What's your illness?

THE TOMB

When the ri-Enewetak arrived on Ujelang in December of 1949, they found a desolate atoll much smaller than their own, covered with brush, and with only a tiny lagoon for fishing. The U.S. Navy left them some tents and canned food, and the ri-Enewetak set about clearing land to cultivate food trees like breadfruit and papaya.

The U.S. eventually built a village on Ujelang with homes, meeting houses, a school, and a church, but the ri-Enewetak found they could not survive on the limited resources of the atoll. Ships with supplies visited infrequently. In 1954, the ri-Enewetak sent a request to the high commissioner of the Trust Territory for more assistance. They received no reply.

In 1956, the Marshall Islanders petitioned the United Nations to stop all nuclear testing in their homeland. Soon after, the U.S. detonated eleven nuclear weapons on Enewetak.

: :

Land occupies a central role in Marshallese culture. This is in part because there is so little of it: about seventy square miles, the size of Washington, D.C., spread across 750,000 square miles of ocean. Over three thousand years people developed ways to thrive on the islands, with fish, breadfruit, pandanus, coconuts, arrowroot, turtles, and birds. They coaxed bare coral islands to flower and flourish. *Los Jardines*, the gardens. In 1529, one of the first Europeans to reach the Marshalls, Spanish sailor Álvaro de Saavedra Cerón, landed on what was probably either Enewetak or Bikini and called it "the gardens."

Kianna Juda Angelo, the woman I met with in Portland, told me that if I'm going to write about the Marshall Islands, I have to visit. I have not yet found a way to make the journey. I have learned from reading, though, that these oases of solid ground in the vast ocean do not lie still, passive beneath one's feet. Land speaks and knows and acts. Each atoll, island, lagoon, and beach conveys lineage and history, sustenance, identity. The preamble to the Marshall Islands constitution lays out the stakes:

> This society has survived, and has withstood the test of time, the impact of other cultures, the devastation of war, and the high price paid for the purposes of international peace and security. All we have and are today as a people, we have received as a sacred heritage which we pledge ourselves to maintain, valuing nothing more dearly than our rightful home on these islands.

: :

When the U.S., USSR and Great Britain signed the Limited Test Ban Treaty in 1963 ending atmospheric testing, the ri-Enewetak requested to go home. But the U.S. military was now using Enewetak for target practice, shooting intercontinental ballistic missiles at the islands from California, testing the ability to fire nuclear warheads at Russia.

In 1972, the U.S. government invited some Enewetak leaders to return to their homeland for a three-day visit. The hereditary chiefs found that three of their islands, Bogairikk, Elugelab, and Teiteriripushi, had completely disappeared. Others had been stripped of vegetation, pockmarked with craters, irradiated, and littered with military machinery.

After their visit, the ri-Enewetak filed a lawsuit against the U.S. Department of Defense to stop testing weapons on the atoll, winning an injunction from a federal judge. In 1973, the U.S. agreed to stop testing and, responding to pressure from the people, announced it would clean up and rehabilitate the atoll so that the ri-Enewetak could go home.

Studies by the U.S. government found radiation "in the soil of the islands, in the sediment at the bottom of the lagoon, in shrubs and trees, and in the flesh of birds and fish." The government considered options for cleaning up the waste, including dumping it into the ocean and shipping it back to the U.S. They picked the least dangerous, least expensive, and least controversial: bury it.

The military used four thousand service members to gather up more than three million cubic feet of radioactive soil and debris and dump it into a nuclear bomb crater on Runit Island, then cover it with a concrete dome eighteen inches thick and nearly four hundred feet across. It took four years to build the Runit Dome. The debris, equal to thirty-four Olympic swimming pools, included plutonium-239, which remains radioactive for hundreds of thousands of years and is dangerous when inhaled.

Six men died during the cleanup, and others later became sick from radiation settling into bones and organs. After the project ended, more radioactive debris washed up, so workers added an antechamber, then a second one.

Then, as the reporter Susanne Rust wrote, they left.

: :

It is possible to see pictures and videos of the Runit Dome online. It looks like a vast, unblinking eyeball staring up at the sky, its grey iris cracked, vines like blood vessels crawling across it. The most dangerous waste—more than four hundred bags of plutonium-contaminated debris—sits at the dome's peak, just below the pupil. A person in a white hazmat suit walks across it, a speck paced by its own dark shadow.

The official U.S. name for the dome is the "crater containment structure." Marshall Islanders call it the Tomb.

BREATHING

In the spring of 2019, U.S. Energy Department scientist Terry Hamilton gave a presentation at Majuro, the capital of the Marshall Islands. He informed the Marshall Islanders that researchers had found high levels of radiation in giant clams near the Tomb.

He also indicated that the dome is vulnerable to rising seas from climate change. The ocean is rising twice as fast in this region of the Pacific than elsewhere in the world, threatening all of the Marshalls. Some predictions say that the Marshall Islands will be mostly underwater by 2100.

Hamilton showed an animation of the Tomb bobbing with the tides as radioactive seawater flooded in and out. "It looked like it was breathing," said James Matayoshi, the mayor of Rongelap Atoll, whose people bore the brunt of the fallout from Bravo, the largest nuclear weapon ever detonated by the U.S.

Hamilton reassured residents that leaks from the Tomb pose no real additional threat because the Enewetak Lagoon is already so contaminated. A scientific study from 1980 concluded that the Tomb contained less than one percent of the plutonium in the lagoon. Plutonium is only a risk when ingested or breathed in, so its presence in water is not an issue, officials say. But contaminated shellfish or groundwater could be. In 2019, researchers from Columbia University found higher levels of radiation on Runit Island than at Chernobyl or Fukushima.

The truth: No site on Earth is like the Tomb. No one knows the risks. "A lot depends on future sea-level rise and how things like storms and seasonal high tides affect the flow of water in and out of the dome," said Ken Buesseler, a scientist at Woods Hole Oceanographic Institution.

In a 2013 report, Hamilton concluded that if the Tomb were in the

U.S., it would be classified as a radioactive waste depository with strict management and monitoring requirements. The Tomb contains no fences and no signs warning people off. It is silent, a staring, concrete eye.

In late 2019, Congress passed a law requiring a thorough study of the risks the Tomb poses to people, the environment, and wildlife. It also required an analysis of the risks climate change poses to the Tomb, though Senate leaders removed the phrase "climate change" from the bill because that would require acknowledging its existence.

: :

In addition to being a poet, Kathy Jetñil-Kijiner co-founded and helps lead a nonprofit called Jo-Jikum that aims to empower youth in the Marshall Islands to "contribute to the survival of their lush, thriving islands."

One of Jo-Jikum's projects, in partnership with the Marshall Islands Conservation Society, is to increase the use of reuseable bags in the Marshalls and discourage single-use plastics. The Marshall Islands struggle to deal with the plastic waste that comes with imported goods. As an island chain with little land area, it can't just find somewhere far away to bury this everlasting garbage. The municipal dump in Majuro is the highest point in the Marshall Islands. Footage shows it rising above an azure sea, with a sea wall to keep it contained. "At this moment, the dump is overflowing," said the Majuro Waste Company's communications official, William Kaisha, Jr., in a video created for the partnership. The waste company is also looking into other ways of managing the continual flow of trash it must handle, including incineration.

Kathy also served as Climate Envoy for the Marshall Islands Ministry of Environment to the United Nations climate negotiations. A UN official interviewed Kathy during the global climate conference held in Madrid in December 2019. Kathy described the critical need for nations to increase their targets for cutting climate pollution in order to avoid disaster for low-lying countries and coastal residents everywhere. "We need all countries to act," she said, "survival...is on the line."

The negotiations ended soon after, with the U.S. and other big polluters blocking even nonbinding language encouraging countries to increase their targets for cutting emissions. It was widely considered the worst outcome in twenty-five years of climate negotiations.

Faced with the seas rising around them, Marshall Islanders are taking matters into their own hands. "Many of our people...want to stay here," then-President Hilda Heine told the *Los Angeles Times* in 2019. "For us, for these people, land is a critical part of our existence. Our culture is based on our land. It is part of us. We cannot think about abandoning the land."

The country is developing a national plan for survival. Kathy told the UN official they are contemplating elevating the islands, or building artificial islands higher above sea level, "to maintain our sovereignty and stay in our own lands," she told him.

He asked her about being a poet as well as a climate envoy. "I need poetry and I need art in my life," she said. "Bearing the weight of being a country that [could be] the first to disappear is incredibly difficult," she said. "There's a grief."

A grief: a specific grief for a specific place. That weight and shape. Kathy called it "a grief," as if it didn't exactly belong to her. I felt it, I imagined, detach from her body and travel into my own chest, along with the weight of the thick cowrie shell necklace that Kathy had given me for a housewarming, assuring me it was very "on trend" in the Marshalls. I clasped the cowrie shells around my neck on a whim as I wrote, to feel connected to her, to the Marshalls, where I've never been.

I imagined this grief expanding and getting bigger, traveling all over the Earth, touching down into every rising and falling chest. How many feel it or know what they felt? Just this specific weight. Weight of a whole people.

BURN

None of the technologies delivering atom-level depictions of benzene can communicate what this molecule does inside the human body.

The name benzene itself carries a history of misunderstanding and confusion. A German chemist in 1934 used the name for a substance he distilled from benzoin, the resin of a tree native to Sumatra known for its sweet, warm scent. The word benzoin comes "via Spanish, Portuguese, or Italian from the Arabic luban jawi, incense of Java,'" even though the tree came from Sumatra. Benzene is called aromatic because of its chemical structure, a name that comes from the sweet smell it generates.

The chemical occurs naturally in oil and gas. It is the scent that lingers after a car engine starts, and it wafts up from the gas pump. It is also among the most widely used industrial chemicals, as a building block for plastics, drugs, pesticides, and dyes. It becomes airborne when heated during all these activities. Igniting tobacco also releases benzene. Because it is so common, low levels of this chemical are found everywhere in the air.

The World Health Organization concludes that there is no safe level of exposure to benzene. When people inhale this chemical, it travels to the center of bones and can damage blood cells forming in the marrow. The damaged, cancerous cells grow fast, travel into the blood, and crowd out the healthy cells the body needs to survive. The specific cancer benzene causes is called acute myeloid leukemia, though it has also been linked to other types of blood cancer. "Acute" means this cancer grows quickly, becomes widespread in bones and blood, and often moves into organs, making it difficult to treat. The survival rate in five years is 27 percent.

This is the most common type of acute leukemia in adults, though it remains an uncommon cancer in the United States, mostly associated with industrial exposures. Radiation is also a risk factor. According to the American Cancer Society: "High-dose radiation exposure (such as being a survivor of an atomic bomb blast or nuclear reactor accident) increases the risk of developing acute myeloid leukemia. Japanese atomic bomb survivors had a greatly increased risk of developing [this cancer]."

Long-term, low-level benzene exposure can damage reproductive organs in women, cause developmental delays in children, and worsen respiratory disease. Texas has the highest rate of benzene pollution in the U.S.—about 1.3 million pounds a year—because of the concentration of chemical plants in the state. The next highest is Louisiana, with 440,000 pounds.

But benzene is only one form of pollution caused by manufacturing plastic and burning fossil fuels. These activities release dozens of other molecules called polycyclic aromatic hydrocarbons, which just means that they are benzene-based, have multiple rings, and are made only of hydrogen and carbon. Molecules like anthracene, pyrene, toluene, and styrene contribute to myriad forms of damage inside the human body: reproductive problems; developmental harm in babies, including low birth weight and poor neurological and immune system development; respiratory illnesses; heart disease; cancer.

The problem with most of these pollutants: They are invisible and have no odor, so it is impossible to know when a person is breathing them.

Producing plastic also creates another invisible form of pollution, the kind that warms the planet. Manufacturing plastic is energy intensive. Cracking apart ethane requires heating it to fifteen hundred degrees and then compressing the ethylene with thousands of Earth atmospheres to form the linked bonds that make polymers. The molecules released from burning all this energy travel high into the atmosphere and stay there for hundreds of years, holding in heat and fevering our sphere.

A report from the Center for International Environmental Law (CIEL) estimates that all the climate-warming pollutants from producing plastic in 2019 equaled 189 large coal-fired power plants. Because plastic production is increasing rapidly, CIEL estimates that number will more than triple, to 615 coal plants worth of pollution by 2050.

: :

The words number and numb come from the same root that meant "to assign, allot, take." Numb refers to the sense of being taken over, overwhelmed. I feel that, studying the numbers. Numb. I sit at a computer screen and feel nothing. These facts don't register in my body.

The shards of the world that catch, that light fire inside me, are people.

In 2009, the scholar Eve Tuck wrote an open letter, an impassioned call to end "damage-centered" inquiries into the suffering of communities and instead practice a "desire-centered" approach. Tuck is Unangax̂, an enrolled member of the Aleut Community of St. Paul Island, Alaska, and Associate Professor of Critical Race and Indigenous Studies at the University of Toronto. Published in the *Harvard Educational Review*, her letter warns against damage-based inquiries serving as "advertisements for power" (a phrase from the scholar Craig Gingrich-Philbrook), by portraying communities only as "broken and conquered." She writes:

> Desire, yes, accounts for the loss and despair, but also the hope, the visions, the wisdom of lived lives and communities. Desire is involved with the *not yet* and, at times, the *not anymore*. In many desire-based texts...there is a ghostly, remnant quality to desire, its existence not contained in the body but still derived of the body. Desire is about longing, about a present that is enriched by both the past and the future. It is integral to our humanness.

I am struck by Tuck's "yes." This is the yes that takes "no" with it, desire as synonymous with life in all its contradictory, disconcerting complexity, the green growth forever escaping the bonds that would contain it:

> Exponentially generative, engaged, engorged, desire is not mere wanting but our informed seeking. Desire is both the part of us that *hankers* for the desired and at the same time the part that *learns* to desire. It is closely tied to, or may even be, our wisdom.

I fear that I have repeated the damage, bearing witness to the pain of these communities. I have not stepped out of grief and caught up to joy. But when I read Tuck's words, I feel that fire, which is love. Love, aching love, for a wounded world that continues to hold and nourish us, love for an albatross obeying its ancient mandate to incubate a future life, even as blows rain down on it, love for the individuals who have blessed me with their friendship, a word that comes from the ancient root for "loving" (**priy-ont*) friend: the present participle form of love, love in action, ongoing.

It is this that comprises hope, the base condition for life, which means "to remain, continue." For love, then:

Miss Jessie in the bright red pickup truck that she requisitioned from the local car dealership for the annual Martin Luther King, Jr. Day parade. She sits beside her daughter, granddaughter, and beaming, infant great-granddaughter. "I started training them early!" she texted. She sent a photo of the Third Place award for her entry in the parade with the note "In it to win IT ALL!"

Miss Jessie, who did not sleep for two nights after the killing of George Floyd because she worried for her twenty-two-year-old grandson, off at a bachelor party with his college friends. She mostly stayed inside during the pandemic. She didn't even go out in the yard for fear the pollution would worsen her asthma and make her more susceptible.

Yukiyo, who sometimes falls into uncontrolled bouts of silent laughter, her mouth open and shoulders shaking. When she first came to Santa Fe, Yukiyo swooned over the adobe architecture, her artist's eye finding the marks of human handiwork in the centuries-old buildings. She ran her hands along the Earth-colored clay, and pressed against the adobe, almost hugging it. Yukiyo, who loves all things sweet. When she took her first bite of a New Mexican sopaipilla, she grinned and closed her eyes, her face shining with delight.

Yukiyo, her eyes looking into me as I fumble for words of an apology that defies speech, or any single lifetime, to deliver.

And Kathy. Now that she has returned home to the Marshall Islands, Kathy's Facebook feed brims with photos of family: Kathy, her mother Hilda, and her daughter Peinam posing together wearing intricate cowrie shell jewelry on Manit Day, a national holiday celebrating Marshallese culture. Kathy in a red tank top and sunglasses, fingers up in a V, standing with youth members of Jo-Jikum, the nonprofit she co-founded.

Kathy in the video production of her poem "Anointed," walking barefoot across the Runit Tomb. It is a poem of desire, of the "*not anymore,*" driven by seeking: "Who remembers you beyond your death?" she asks, addressing the island destroyed by nuclear blasts and turned into a plutonium dump. "Who would have us forget?"

The poem arrives at frustration, at the inability to heal the loss of an entire island, and all it meant, and at the relentless continuing damage caused by humans intent on setting the planet on fire. The poem rises in a crescendo to its final questions. Kathy stands at the highest point of the Tomb, in the center of the concrete eye, and spits out these words with a fury that vibrates the air:

Who gave *them* this power?

Who anointed *them* with the power to burn?

LEGACY: "THAT'S YOURS"

By the time we arrived in Birmingham, Alabama, the night before the Honda plant tour, I felt strange, dislocated from everything familiar, after many days on the road. I dropped Cobalt and Aster off at a hotel and wound through the traffic-clogged streets of suburban Birmingham at rush hour, looking for a place to buy dinner. I'd been thinking about taking this trip to the Honda Odyssey plant for five years. I'd been talking about it, too. Telling people how I planned to bring my car part cross country someday and present it to the people at Honda. I thought of it as a gesture, a bit outrageous, to bring a large piece of plastic garbage two thousand five hundred miles across the country and ask the company responsible for its manufacture, its birth, to also take responsibility for its neverending death.

I thought it would be a symbol: What if we in industrial societies took all our garbage this seriously? We couldn't function, and neither could the companies who make this stuff. They would have to change, to stop profiting by leaving the refuse—the pollution, the waste—to be absorbed by other bodies.

As Cobalt and I started thinking about this road trip through the South, one of us at some point came up with the idea of Aster delivering the car part to the factory. We thought it might make the adults more receptive. We talked to her about it, and what she might say. We even, I'm ashamed to say, coached her.

I learned online that the plant offers free tours on Wednesdays and Thursdays. They required advance reservations and only people over twelve could attend. That didn't deter us. We figured we'd bring Aster and the car part in, and then I would take the tour alone. I had to fill out a form online and agree to a series of rules. I could not bring a camera or a phone or a purse or a bag of any kind, nothing but paper and pencil. I had to list the names of everyone accompanying me. I had to wear pants and closed-toe and closed-heel shoes.

The Honda Lincoln Assembly Plant is located in a largely rural area between the Talledega Superspeedway and the Coosa River, just outside the town of Lincoln, Alabama, and forty miles from Birmingham. We planned to leave early, in time to make the 9 a.m. tour. By the time I returned to the hotel room with a store bought dinner, I felt exhausted. I also felt nervous. The tour the next day felt like the culmination of so many years of obsession. I had no idea, no sense of what might happen.

Aster did not want to go. She was also tired from days on the road, hotel after hotel, watching movies in the car while I met with various community leaders. It wasn't exactly a kid-centric trip, except that we tried to find a place with a pool every night as relief from the southern heat. As she got under the covers that night, she said she felt too afraid to go to the factory. "I get stage fright you know," she told me.

I felt a wave of disappointment, also, a tinge of desperation. This was not going according to plan. I started to try and convince her, but my adult brain intervened. "You cannot," I told myself, "force an eight-year-old to do something she doesn't want to." I kissed her, told her that was fine, that she and Cobalt could just meet me after the tour.

I stayed up most of the night, scripting in my head what I would say and do at the plant. In the morning, Cobalt got up with my alarm and woke Aster. "We're coming with you," Cobalt said. "We'll wait out front."

I don't remember the drive to the plant, I was too worried about being lost, being late. We pulled up to a blocky beige building that said Honda in red letters, fronted by a sprawling parking lot. I got out, and Aster climbed out after me, looking still a little sleepy, the car part draped around her neck, holding her stuffed leopard "Noon" in her arms. "I'm coming with you," she said, full of nonchalance.

We walked into the lobby. Cobalt captured all that followed on the GoPro they held surreptitiously at their waist. Natalie, our tour guide, approached, wearing the regulation Honda uniform of white pants and a white button-down shirt and carrying a clipboard. She was a tall woman, a head taller than I am, and, I learned later, a retired teacher. She had that teacher's air of no nonsense about her. "You can't bring that," she said, by way of greeting, nodding at my bag and holding out the clipboard for me to sign.

Aster stepped in front of me, waving the car part, "I have a question," she said. "Could you re-use this?"

We had not discussed that she would say this. In fact, nothing she said during this conversation had anything to do with what we rehearsed. I followed her lead, explaining, "We found this in our front yard, and it turns out it comes from a Honda Odyssey." I kept my voice light and friendly.

"Well, I'm sure we wouldn't be able to use *that*," said Natalie, drawing out the word "that" with her Alabama drawl.

"We're here from Oregon," I continued, "We can't recycle it there, and we don't want to just throw it away."

"I don't know where *we* could recycle it, is the deal," said Natalie, lifting her shoulders high in a shrug.

"You could just turn it into another car part," suggested Aster.

"We can't," said Natalie, with finality, and a slight edge of impatience creeping into her voice. "There's no way to do that."

"We heard you're a zero-waste plant," I persisted. "So we thought you could take responsibility for it."

Natalie shook her head. "Don't think so. Just because, you know..." she waved her hand helplessly. "We recycle everything we produce *here*, but we don't bring in things from offsite."

"But that's pretty cool that you held on to that," she told Aster somewhat patronizingly, smiling at her. I realized, in a flash, why Aster's presence was so important to me. She gave me a role; she made me recognizable. I was not just some potentially unhinged person carrying around a large piece of plastic garbage. I was a parent—probably, in their eyes, a mother—trying to do my best for the child in my life.

"It's from a first-generation Odyssey," I said. "The very first one."

"A first-generation Odyssey, huh?" said Natalie. "Well, that's gotta be valuable."

"Maybe you could put it up for display somewhere," said Aster, excited by this idea.

"We don't have room to display anything," replied Natalie, again with finality.

Aster turned her whole body and took a slow pan around the large lobby. "Actually," she said, "you have lots of room to display it." But Natalie had already turned away from her and was giving Cobalt ideas for where to get breakfast while they waited for me. Aster perked up at the sound of Waffle House and turned toward Cobalt, and the camera lens, to nod vigorously. Then she turned back to Natalie. "You could actually put this anywhere," she said, swinging the car part in front of her.

Natalie smiled and leaned toward her, a cheerful, hard edge of non-negotiation in her voice. "They won't let me take anything like that here," she said.

"That's yours."

NOTES

Word etymologies throughout the book come from Douglas Harper's wonderful online etymology dictionary (etymonline.com), and from the *Oxford English Dictionary.*

GIFT: THE THING (3-6)

Epigraph: Solnit, *Recollections of My Nonexistence: A Memoir*, 126.

"I am the *no* and the *yes*": Sobelman, *The Tulip Sacrament*, 8 (italics in the original).

Some worlds have ended over and over: Yusoff, *A Billion Black Anthropocenes or None*, 83-121.

"See to it that Americans are never satisfied": Meikle, *American Plastic: A Cultural History*, 176.

I. THE THING

WORK (9-10)

"The eyes of a prophet": Rota, "The Lost Café," *Los Alamos Science*.

"It is a totally difference scheme": Ulam, "Postscript to Adventures," 305.

STORY (11-13)

Long's disappearance: Philip Bassett, email correspondence with the author, shared excerpts from the journal of Joseph F. Long, January 2012.

All he could see were flames: Messimer, *In the Hands of Fate*, 114-122; Brown, *Suez to Singapore*, 459-479; Creed, *PBY*, 88-89; Knott, *Black Cat Raiders of WWII*, 25-30; Christman, "Action Report of Commanding Officer."

ZERO (14)

History of zero: Seife, *Zero*, 29-36; Kaplan, *The Nothing That Is*, 14-27.

NOTHING (15)

***Doorbell* and *Mercy floated toward me*:** Notley, "Desert Doorbell," 81.

"I am the *no* and the *yes*": Sobelman, *The Tulip Sacrament*, 8.

The no-thing desire toward being: Barad, "After the End of the World" 524-550.

THE IMPURITY (16-18)

William Perkin and the discovery of mauve: Colorants Industry History, "William H. Perkin"; Royal Society of Chemistry, "Sir William Perkin"; Garfield, *Mauve*, 19-23; Filarowski, "Perkin's Mauve," 850-855.

History of Tyrian purple: Morris and Travis, "A History of the International Dyestuff Industry"; Dr. Richard M. Podhajny, "History, Shellfish, Royalty, and the Color Purple"; Sandberg, *The Red Dyes*, 17-40.

WORK (19-20)

My understanding of emotion and emotional work and labor is informed by Hochschild, *The Managed Heart* (24-34, 68-75, 131-135, 147, 155-170, 187-197). See also: Julie Beck, "The Concept Creep of 'Emotional Labor'," *The Atlantic*, November 26, 2018: https://www.theatlantic.com.

THE MARINER (21-25)

"Man stares": Heidegger, "The Thing," 166.

***The Rime of the Ancient Mariner* and the context of its composition:** Coleridge, "The Rime of the Ancient Mariner"; Safina, *Eye of the Albatross*, 25-28; Holmes, *Coleridge: Early Visions*, 171-172.

Heidegger looked around and saw corpses: Heidegger, "The Question Concerning Technology," 3-35.

THE CURTAIN (26-28)

Ulam finding out war had broken out: Ulam, *Adventures of a Mathematician*, 33, 58-122.

Ulam family history and experience leading to the war: Ulam, *Understanding the Cold War*, 5, 20-21, 27-28, 81-95; Mycielski, "Measurable Cardinals," 107.

FRAGMENT (31-34)

Book on ocean trash: Ebbesmeyer and Scigliano, *Flotsametrics*.

Unimog: Patrick E. George, "How Unimogs Work," *How Stuff Works*, February 28, 2011, www.howstuffworks.com.

THE RING (35-36)

Ouroboros vision: Rocke, *Image and Reality*, 194; Kekulé, "August Kekulé Speech Berlin City Hall, 1890."

Puerperal fever: Nuland, *The Doctors' Plague.*

***Females and Their Diseases*:** Loudon, *Death in Childbirth*, 56; Meigs, *Females and Their Diseases*, 596; Meigs, *On the Nature, Signs and Treatment of Childbed Fever*, 104.

LOSS (37-41)

Telegram: Lance Christman, personal collection.

Christman's experience of being shot down: Brown, *Suez to Singapore,* 475-479; Christman, "Action Report of Commanding Officer."

Plexiglas in airplanes: Meikle, *American Plastic*, 86-88; Dorny, *US Navy PBY Catalina Units of the Pacific War*, 8-9.

Mt. Angel and Old Believers: Garcia, "Latinos in Oregon"; Binus, "Russian Old Believers"; Morris, Morris, and Osipovich, "Old Believers."

Kalapuya: Esther Stutzman, "Indigenous Storytelling: Kalapuya Creation Story," The Archeology Channel, August 30, 2001, https://www.archaeologychannel.org/audio-guide/indigenous-storytelling-kalapuya-creation-story; Donny Morrison, "The Future of the Kalapuya Story," *The Daily Emerald*, January 28, 2019, https://www.dailyemerald.com/news/the-future-of-the-kalapuya-story/article_61a1fb7a-22ab-11e9-bf8f-9be233937004.html; Don Macnaughton, "Kalapuya: Native Americans of the Willamette Valley, Oregon," Lane Community College Library, June 4, 2020, https://libraryguides.lanecc.edu/kalapuya; Confederated Tribes of Grand Ronde, "Our Story," https://www.grandronde.org/history-culture/history/our-story/.

THE ALBATROSS (42-44)

Susan Middleton on photographing albatross: Liittschwager and Middleton, *Archipelago*, 34, 201-203; Susan Middleton, interviews with the author, February and March 2011.

Tracking albatross: "Tagging of Pelagic Predators," *Tagging of Pelagic Predators*, http://www.gtopp.org; Perras and Nebel, "Satellite Telemetry," 4.

Appearance of albatross: Richard Ellis, quoted in Safina, *Eye of the Albatross*, 2-3; *Eye of the Albatross*, 24; Liittschwager and Middleton, *Archipelago,* 34.

MAUVE (45-47)

Empress Eugénie: Dolan, "The Empress's New Clothes," 22-28; Seward, *Eugénie*; McQueen, *Empress Eugénie and the Arts*, 127.

Crinoline cage: Nead, "The Crinoline Cage."

Mauve: Morris and Travis, "A History of the International Dyestuff Industry"; Brunello, *The Art of Dyeing*, 38.

"Macho" Prussian colors: Dolan, "The Empress's New Clothes," 26.

Lust launched industrial chemistry: Garfield, *Mauve,* 44-86.

THE PHOTOGRAPH (48-50)

***I die therefore I am*:** Goldman, "The Dispossessions," 139.

Story of Shed Bird and interview: Liittschwager and Middleton, *Archipelago,* 205; Susan Middleton, personal interviews with the author, March 2011, January 2013, and July 2013.

GYRE (51-53)

One of my first encounters with gyres, Dr. Ebbesmeyer, and the notion of marine plastic pollution was Donovan Hohn's 2006 article in *Harper's Magazine* entitled "Moby-Duck." His book of the same name helped spark my own plastic odyssey.

On gyres: Ebbesmeyer and Scigliano, *Flotsametrics*, 157, 186-207.

Sources of ocean trash: Bengali, "Your Trash Is Suffocating this Indonesian Village"; Jambeck et al., "Plastic Waste Inputs From Land Into the Ocean," 768-771.

Hawai'i Wildlife Fund clean ups: "Hawai'i Island Marine Debris Removal Project," *Hawaii Wildlife Fund*, http://wildhawaii.org.

WHITE WHALE (54-56)

Lugworm: Barnes et al., "Accumulation and Fragmentation," 1985-1998; Browne et al., "Microplastic Moves Pollutants," 2388-2392, https://doi.org/10.1016/j.cub.2013.10.012; Wright et al., "Microplastic Ingestion," R1031-R1033, https://doi.org/10.1016/j.cub.2013.10.068.

DISAPPEARANCE (57-59)

Ulam secures work at Los Alamos: Ulam, *Adventures of a Mathematician*, 87, 117-118, 122-124, 136-138, 141-143.

INFINITY (60-62)

Kekulé's grief: Rocke, *Image and Reality*, 199.

Understanding of molecular structure and benzene ring: Spector, "Nanoaesthetics," 1-29; Rocke, *Image and Reality*, 1-205, 211, 297.

Linus Pauling and resonance: Hargittai, *Judging Edward Teller*, 114-115.

Infinity, plastic's first trademark symbol: Meikle, *American Plastic*, 1, 31-62.

GUSH (63-68)

***Matter is pitiful; form is terrible*:** Robertson, "7.5 Minute Talk for Eva Hesse," 43.

***The Rime of the Ancient Mariner*:** Coleridge, "Rime of the Ancient Mariner."

Gush: Heidegger, "The Thing," 173. In a more recent translation of "The Thing," translator Andrew J. Mitchell gives *giessen* as "pour." See also: Heidegger, *Bremen and Freiberg Lectures*, 11; "Gush," *Vocabulary*, accessed October 26, 2019, https://www.vocabulary.com.

Odysseus's hunger: Homer, *The Odyssey*, 201, 215, 402.

Albatross with toothbrush: Safina, *Eye of the Albatross,* 277-278.

Albatross as a mark of guilt: I am indebted to Safina for this reading of Coleridge.

DESIRE (69-71)

Photograph of benzene molecule: Browne, "A Pervasive Molecule."

Kekulé may have made up the ouroboros story: Rocke, *Image and Reality,* 198-199; Larsen, "Kekulé's Benzolfest Speech," 184.

The fiction of sight in depicting molecules: Specter, "Nanoaesthetics," 1-29.

Such an image "gives the feeling one could touch the atoms": Goodsell, "Fact and Fantasy in Nanotech Imagery," 52-57.

"The experience...sent chills through my body": Barad, *Meeting the Universe Halfway*, 37.

Desire and molecular imagery: Tami Spector, email correspondence with the author, September 2013.

Chemical reaction imagery: Berkeley Lab, "Atom by Atom, Bond by Bond, a Chemical Reaction Caught in the Act."

"All images are after": Retallack, *The Poethical Wager*, 10.

RESURRECTION (76-78)

Elwyn Christman crash landing and return to U.S. Naval headquarters: Crocker, *Black Cats and Dumbos*, 235; Messimer, *In the Hands of Fate*; Brown, *Suez to Singapore,* 479; Dawley, "Bombing Attack at Jolo, Sulu"; Lance Christman, Personal collection of Elwyn Christman letters.

Moro encounters with U.S. troops: Arnold, *The Moro War*, 1-7.

GARBAGE (79-83)

Shed Bird picture: Middleton and Liittschwager, "Hawaii's Outer Kingdom," 70; Susan Middleton, personal interview with the author, February 11, 2011.

Origin of WWII-era piece of plastic inside Shed Bird: Louis Dorny, email correspondence with the author, March 3, 2011; Ebbesmeyer and Scigliano, *Flotsametrics*, 212; PBY online discussion group, retrieved from Yahoo! Groups exchange, October 2005; "Notice," VPNAVY, October 9, 2005: https://www.vpnavy.org/vp101_notice.html.

***500 lb. eggs with T.N.T. guts*:** Lance Christman, Personal collection, Elwyn Christman letter.

"How the egg of a bird is crystalline": Smith, "On Offer and On Reflection," 154.

"Lots of loose ends": Louis Dorny, email correspondence with the author, March 3, 2011; PBY online discussion group, retrieved from Yahoo! Groups exchange, November 1, 2005.

I first encountered the piece of plastic in the *Los Angeles Times*: Weiss, "Altered Oceans."

"Any kind of knowledge can be a prescription against despair": Rankine, *Don't Let Me Be Lonely*, 55.

Oldest piece of plastic from the ocean: Ebbesmeyer and Scigliano, *Flotsametrics*, 212; Curtis Ebbesmeyer, email correspondence with the author, February 21, 2011.

The piece of plastic has been lost: Susan Middleton, email correspondence with the author, January 22, 2013.

REFUSE (84-86)

Invention of Bakelite: American Chemical Society, "Leo Hendrick Baekeland and the Invention of Bakelite"; Kaufmann, "Leo H. Baekeland."

Albumen prints help drive the slaughter of albatross: James M. Reilly, "The History, Technique and Structure of Albumen Prints," *The Albumen & Salted Paper Book*, 1980, https://cool.culturalheritage.org/albumen/library/monographs/reilly/.

ADVENTURE (87-92)

***eely things water weedy things*:** Notley, *Close to Me & Closer*, 22.

Ulam getting a feel for physics: Ulam, *Adventures of a Mathematician*, 147-148.

Edward Teller led the work on the hydrogen bomb: Ulam, *Adventures of a Mathematician* 149-151; Rhodes, *The Making of the Atomic Bomb*, 370; Rhodes, *Dark Sun*, 246–249.

Teller history: Hargittai, *Judging Edward Teller*, 61-69, 128-134.

Teller experiment provides evidence for Pauling's resonance theory: Hargittai, *Judging Edward Teller*, 114-116; Nordheim et al., "Note on the Ultraviolet Absorption Systems of Benzene Vapor."

Terry Bristol tour of Linus Pauling house in Portland: Terry Bristol, meeting with the author, September 3, 2020.

Linus Pauling history: Paradowski, "Linus Pauling: American Scientist."

Pauling-Teller debate: Melinda Gormley and Melissae Fellet, "The Pauling-Teller Debate: A Tangle of Expertise and Values," *Issues in Science and Technology* 31, no. 4 (2015): 78-82; "Fallout and Disarmament: A Debate between Linus Pauling and Edward Teller," *Daedalus* 87, no. 2, (1958): 147-163, https://www.jstor.org/stable/20026443.

Enrico Fermi suggests the idea of the hydrogen bomb: Rhodes, *The Making of the Atomic Bomb*, 374; Rhodes, *Dark Sun*, 248.

ODYSSEY (93-95)

The legend of the Honda Odyssey: Honda, "Developing a car with a roomy interior," https://global.honda/heritage/episodes/1994odyssey.html; Maynard, *The End of Detroit*, locs. 1415, 1421, 1437, 1462, 1473.

Kunimichi Odagaki: Maynard, *The End of Detroit*, 1415.

"You will lack for nothing": "2014 Honda Odyssey is the world's first minivan that sucks on purpose," *New York Daily News*, August 23, 2013, https://www.nydailynews.com/autos/latest-reviews/2014-honda-odyssey-world-minivan-sucks-purpose-article-1.1435102.

II. REFUSE

REFUSAL (99-102)

Earliest human remains found in Summer Lake region: Jenkins et al., "Clovis Age Western Stemmed Projectile Points," 223-228.

THE COMPUTERS (103-106)

Struggle to design the thermonuclear bomb: Rhodes, *Dark Sun*, 249, 250-251, 370-374, 400-460; Rhodes, *Making of the Atomic Bomb*, 77, 755-758, 773;

Ulam, *Adventures of a Mathematician*, 209, 213, 214, 215, 218, 310.

Women as computers, and computer programmers: Howes and Herzenberg, *Their Day in the Sun*, 93-110; Light, "When Computers Were Women," 461-471, see note 20 for quotation: "constant alertness..."; Fritz, "The Women of ENIAC," 13-14.

The computers provided a diversion: Ulam, *Adventures of a Mathematician*, 218.

A "room with girls in it": Howes and Herzenberg, *Their Day in the Sun*, 107.

Legacy of women software pioneers: Fritz, "The Women of ENIAC," 17; Steve Lohr, "Frances E. Holberton, 84, Early Computer Programmer," *The New York Times*, December 17, 2001; Steve Lohr, "Jean Bartik, Software Pioneer, Dies at 86," *The New York Times*, April 7, 2011.

Women left out of publicity about the first computer: Light, "When Computers Were Women," 472-477; T.R. Kennedy Jr., "Electronic Computer Flashes Answers, May Speed Engineering," *The New York Times,* February 15, 1946.

THE CONSUMERS (107-110)

Women as substitute workers like plastic is substitute material: War Department, "You're Going to Employ Women"; Light, "When Computers Were Women," 455-483.

Promotion of plastic for wartime use: Meikle, *American Plastic*, 1, 160-161, 167, 180; Rogers, *Gone Tomorrow*, 120; General Electric, *How Plastics Solved War Problems*; Fenichell, *Plastic: Synthetic Century*, 203.

Promoting plastic to postwar consumers and rise of the advertising industry: Rogers, *Gone Tomorrow*, 120-124; Meikle, *American Plastic*, 265; Wilson and Lande, "Feeling Capitalism," 275-278.

Creating disposable uses for plastic: Meikle, *American Plastic*, 265-266.

Packaging absorbs 40% of plastic and accounts for the majority of plastic waste: Laura Parker, "Fast facts about plastic pollution," *National Geographic,* December 20, 2018, www.nationalgeographic.com; Hannah Ritchie and Max Roser, "Plastic Pollution," Our World in Data, September 2018, ourworldindata.org.

Certain organisms may be eating plastic: Santoro, "These Tiny Microbes are Munching Away at Plastic Waste in the Ocean."

Essentially all plastic ever created still exists: Geyer, Jambeck, and Law, "Production, Use, and Fate of All Plastics Ever Made," 1-5; Sarah Knapton,

"Plastic weighing equivalent of one billion elephants has been made since 1950s and most is now landfill," *The Telegraph*, July 29, 2017; Ari Phillips, "Plastics Pollution on the Rise," Environmental Integrity Project, September 5, 2019, environmentalintegrity.org, 1.

LAMENT (111-115)

Rodents eating albatross alive on nests: Safina, *Eye of the Albatross*, 270-271.

Catalogue of human abuses of albatross: Safina, *Eye of the Albatross*, 80-150, 250.

U.S. Navy PBY crews' interactions with albatross: Crocker, *Black Cats and Dumbos*, 50-52.

Teens on O'ahu kill nesting albatross: Nelson Daranciang, "Teen receives 45-day sentence in Kaena Point albatross slaying," *Honolulu Star Advertiser*, July 6, 2017; Jennifer Sinco Kelleher, "Hawaii prep school grad gets 45 days jail for bird deaths," *Associated Press*, July 6, 2017; Brittany Lyte, "Prep school teens were accused of massacring protected birds. Did they get off too easy?" *The Washington Post*, July 5, 2017.

Fishing and albatross deaths: Sarah Bladen, "The tale of the albatross and the algorithm," *Birdlife International*, January 27, 2019; Danielle Hall; "Saving Albatross Lives with Bird Scaring Lines," *Smithsonian*, August 2017; Safina, *Albatross*, 188-193.

Another image of the albatross haunts me: Photograph by Graham Robertson in Safina, *Eye of the Albatross*, 188.

Humans and albatross ingest plastic: Anna Turns, "Saving the Albatross: 'The war is against plastic and they are casualties on the frontline,'" *The Guardian*, March 12, 2018; Laura Parker, "Nearly Every Seabird on Earth Is Eating Plastic," *National Geographic*, September 2, 2015; Sarah Gibbens, "You eat thousands of bits of plastic every year," *National Geographic*, June 5, 2019, www.nationalgeographic.com; Emma Stoye, "Humans consuming thousands of microplastic particles in their food every year," *Chemistry World*, June 10, 2019; Antonio Ragusa, et al. "Plasticenta: First evidence of microplastics in human placenta."

Plastic "rain" found everywhere on Earth: Brahney et al., "Plastic Rain in Protected Areas of the United States," 1257-1260; John Schwartz, "Where's Airborne Plastic? Everywhere, Scientists Find," *The New York Times*, June 11, 2020.

LOSS (123-126)

Ida Noddack: Santos, "A Tale of Oblivion," 373-389; Habashi, "Ida Noddack," 215-217; Sime, *Lise Meitner*, 273; Pearson, "On the Belated Discovery of Fission," 40-45; Fermi, "Possible Production of Elements," 898-899; Howes and Herzenberg, *Their Day in the Sun*, 30.

"The reason for our blindness is not clear": Segre, *Enrico Fermi*, 76.

Lise Meitner: Sime, *Lise Meitner*, 6-9, 57-70, 162-183, 210, 228, 233-237, 260-265, 291, 305, 307-309; Watkins, "Lise Meitner, the Foiled Nobelist," 163-164, 173, 178-179, 180-181; Lise Meitner, "Looking Back," *Bulletin of Atomic Scientists* 20, no. 9 (1964): 2-7.

"I will have nothing to do with a bomb!": Sime, *Lise Meitner*, 305.

Françoise Ulam as Stan Ulam's "human word processor": Ulam, "Postscript to Adventures," 305-308.

Françoise Ulam and the bomb: Claire Ulam Weiner, interview with the author, December 27, 2013.

Grief is a god, writes the poet Alice Notley, "as in possession": Alice Notley, "II — The Person That You Were Will Be Replaced," *Mysteries of Small Houses* (New York: Penguin Books, 1998), 78-79.

GRIEF (127-129)

Jiro Horikoshi childhood: Horikoshi, *Eagles of Mitsubishi*, 8-12.

WWI treaty limiting armaments on ships led to Japan's focus on planes: Horikoshi, *Eagles of Mitsubishi*, 14; United States Department of State, "The Washington Naval Conference, 1921-1922."

"Did you see that cute tracer?": Horikoshi, *Eagles of Mitsubishi*, 14.

Type 96 airplane in China: Horikoshi, *Eagles of Mitsubishi*, 14-25; D'Angina, *Mitsubishi A6M Zero*, 8-10

Nanking massacre: Mitter, *Forgotten Ally*, 124-144; Yoshida, *The Making of the "Rape of Nanking,"* 11-42.

"A way forward is to ask a different kind of question": Simon Han, "The Impossible Task of Remembering the Nanking Massacre," *The Atlantic*, December 17, 2017.

Horikoshi's death: Horikoshi, *Eagles of Mitsubishi*, 26.

Grief and the lungs: Lehrer, "Anger, Stress, Dysregulation," 833-834; Emma Suttie, "Grief and the Lungs," *Chinese Medicine Living*, accessed August 19, 2018, www.chinesemedicineliving.com.

ZERO (133-135)

Designing the Zero airplane: Horikoshi, *Eagles of Mitsubishi*, 3-8, 25-28, 30, 40-41, 44, 58, 70, 95.

Origins of zero: Sarah Gibbens, "Ancient Text Provides New Clues to the Origins of Zero," *National Geographic*, September 16, 2017; Kaplan, *The Nothing That Is*, 4-36, 80-84; Robert Kaplan, "What Is the Origin of Zero?" *Scientific American*, February 28, 2000; Seife, *Zero*, 1-73.

RADAR (136-138)

The discovery and production of polyethylene: Meikle, *American Plastic*, 158; Fenichell, *Plastic: Synthetic Century*, 200-202; Freinkel, *Plastic: A Toxic Love Story*, 60-61; Allen, *Studies in Innovation*, 20-22, 26-27; Swallow, "A History of Polythene," 1-4.

Polyethylene and radar: Fenichell, *Plastic: Synthetic Century*, 202-203; Allen, *Studies in Innovation,* 26-27; Louis Brown, *A Radar History of World War II*, 39-40; Swallow, "A History of Polythene," 4-8.

PBYs first U.S. Navy planes with radar: Brown, *Radar History*, 162, 364.

Radar and World War II brutality: Buderi, *The Invention that Changed the World*, 80-97, 172-230, 237-240; Brown, *Radar History*, 322, 323, 422-426; Kate Connolly, "Panel rethinks death toll from Dresden raids," *The Guardian*, October 2, 2008.

Curtis LeMay's war criminal comment: Buderi, *The Invention that Changed the World*, 237-240; Brown, *Radar History*, 422-426; Matthew Gault, "God No, the U.S. Air Force Doesn't Need Another Curtis LeMay," *Medium*, April 7, 2015, https://medium.com/war-is-boring/god-no-the-u-s-air-force-doesn-t-need-another-curtis-lemay-37de19c11652.

JOB (139-142)

The Zero took the Allies by surprise: Okumiya, Horikoshi, and Caidin, *Zero!*, 102-103; D'Angina, *Mitsubishi A6M Zero*, 5; Yoshimura, *Zero Fighter*, 131-136; Stephan Wilkinson, "Myth of the Zero," *Aviation History Magazine*, December 1, 2017, https://www.historynet.com/myth-of-the-zero.htm.

"The days of the frightening dreams": Horikoshi, *Eagles of Mitsubishi*, IX.

Horikoshi during World War II: Horikoshi, *Eagles of Mitsubishi*, 64-64, 130, 142-145; Yoshimura, *Zero Fighter*, 139, 175, 191-192.

Battle of Midway: "The Battle of Midway," National World War II Museum, accessed October 18, 2018, www.nationalww2museum.org, see: Albatross

Photo on Burning Midway; Andrew Lambert, "The Battle of Midway," BBC History, February 17, 2011.

Kamikazes use Zero airplanes: Yoshimura, *Zero Fighter*, 185-200; D'Angina, *Mitsubishi A6M Zero*, 58; Nijboer, *Seafire vs. A6M Zero*, 62-63.

Horikoshi on kamikazes: Horikoshi, *Eagles of Mitsubishi*, 146-147.

Horikoshi at the end of WWII: Horikoshi, *Eagles of Mitsubishi*, 148-149; Okumiya, Horikoshi, and Caidin, *Zero!*, 889-904.

Horikoshi family: Jeff Penberthy, "Plane Designer Recalls Days of Zero's Success," *Los Angeles Times*, December 14, 1972.

WIND (143-145)

***The Wind Rises*:** Rebecca Keegan, "'The Wind Rises': Hayao Miyazaki's new film stirs controversy," *Los Angeles Times,* August 15, 2003; Jeremy Blum, "Animation legend Hayao Miyazaki under attack in Japan for anti-war film," *South China Morning Post*, August 13, 2013; Robbie Collin, "Hayao Miyazaki interview: 'I think the peaceful time that we are living in is coming to an end'," *London Telegraph*, May 9, 2014; Napier, *Miyazakiworld*, 382-517.

***The wind is rising!...*:** Paul Valéry, "The Graveyard by the Sea," trans. C. Day Lewis, https://www.poemhunter.com/poem/the-graveyard-by-the-sea/.

TO LIVE (146-150)

***How Do You Live?*:** Jennifer Sherman, "Hayao Miyazaki Directs Next Film Without Deadline for Finishing," *Anime News Network*, March 12, 2019; Hoai-Tran Bui, "Hayao Miyazaki's Next Movie is Still 3 to 4 Years Away," *Slash Film*, August 8, 2018.

Nationalism in Japan: Saaler, "Nationalism and History in Contemporary Japan"; "Asia's Second World War Ghosts," *The Economist*, August 15, 2015; Sheila A. Smith, "Episode 8: The Character of Japanese Nationalism," December 6, 2017, in *Nationalism, Japan, and a Changing Asia*, podcast, MP3 audio, 39:45, https://www.cfr.org/podcast-series/nationalism-japan-and-changing-asia.

Nobusuke Kishi: Reiji Yoshida, "Formed in childhood, roots of Abe's conservatism go deep," *The Japan Times*, December 26, 2012; "Asia's Second-World-War Ghosts," *The Economist*, August 15, 2015.

Grandfather's WWII service: Personal correspondence with the author's aunt, January 11, 2019.

Yukiyo Kawano's art, and her grandfather's atomic bomb testimony: Kawano, "Active Intransitive," 29-30.

Pained entanglements: My understanding of responsibility and entanglement is influenced by Karen Barad. The words "responsible" and "refusal" both come from the same root Heidegger used for the word "gush"—*gheu*, a flow. "Response" means to bathe with another in the flow. "Refuse" means to stop the flow, to pour it back.

THE THINKER (151-156)

Claire Ulam Weiner on her father, Stan Ulam: Interview with the author, December 27, 2013.

Alex Ulam in Lviv: Alex Ulam, "Lviv's, and a Family's, Stories in Architecture," *The New York Times*, October 17, 2013

***From Cardinals to Chaos*:** Los Alamos Science, "Special Issue: Stanislaw Ulam," 1; Cooper, Eckhardt, and Shera, eds., *From Cardinals to Chaos*.

Mary Tsingou-Menzel: Dauxois, "Fermi, Pasta, Ulam, and a Mysterious Lady," 55-57; Roger Snodgrass, "A not-so-mysterious woman," *Los Alamos Monitor*, August 3, 2008.

View of Omega Bridge: Email correspondence with the author's father, September 24, 2015.

"All my father does is think, think, think!": Los Alamos Science, "Special Issue: Stanislaw Ulam," 1; Cooper, Eckhardt, and Shera, eds., *From Cardinals to Chaos*, 320.

SEED (157-158)

"To be entangled...": Barad, *Meeting the Universe Halfway*, ix.

HERE (159-162)

Artist who makes jewelry out of nurdles: Zoe Donaldson, "This Gorgeous Jewelry is Made from Plastic Collected from the Beach," www.oprah.com.

HAUNT (163 - 166)

Imagining the Fukushima disaster: Richard Lloyd Parry, "The school beneath the wave: the unimaginable tragedy of Japan's tsunami," *The Guardian*, August 24, 2017.

Hauntology: I'm indebted to Karen Barad for the concepts regarding this word coined by Jacques Derrida. For Barad's conceptualization see: Barad, *Meeting the Universe Halfway*; Barad, "Quantum Entanglements and Hauntological Relations," 240-268; Barad, "After the End of the World,"

539. See also: Tom Whyman, "The ghosts of our lives: From communism to dubstep, our politics and culture have been haunted by the spectres of futures that never came to pass," *New Statesman*.

BACKYARD (167-169)

Heidegger lecture "The Thing": Heidegger, *The Question Concerning Technology and Other Essays*, ix; Harman, *Heidegger Explained*, 12; Safranski, *Martin Heidegger*, 390-392.

Heidegger banned from teaching because of joining the Nazis: Safranski, *Martin Heidegger*, 336-342; Harman, *Heidegger Explained*, 11.

Ulam on his friends and family murdered by Nazis: Stanislaw M. Ulam, *Adventures of a Mathematician*, 27, 37, 40, 43, 172-173.

Heidegger's Freiberg lectures: Harman, *Heidegger Explained*, 1, 2; Heidegger, *Introduction to Metaphysics*, 39-45.

"Backyard" delivery weapon: Serber, *The Los Alamos Primer*, 4, quoted in: Rhodes, *Dark Sun*, 253.

BUTCH (173-178)

Elwyn Christman, his son, and his death: Raymond Hurlbut, interview with the author, 15 August 2011; Lance Christman, interview with the author, 13 July 2011; Lance Christman, personal collection, photos and letters.

MIKE (180-184)

Revealing the inner workings of the hydrogen bomb: Morland, *The Secret That Exploded*, 24-228; Rhodes, *The Making of the Atomic Bomb*, 770-780; Rhodes, *Dark Sun*, 462-463.

Plastic crucial to the bomb's design: Rhodes, *Dark Sun*, 492-506; Richard Rhodes, email correspondence with the author, February 2014; Nick Baumann, "Did America Forget How to Make the H-bomb?" *Mother Jones*, May 1, 2009; Los Alamos National Laboratory, "Fogbank, Lost Knowledge Regained," 20-21.

Polyethylene: Fenichell, *Plastic: Synthetic Century*, 236; Freinkel, *Plastic: A Toxic Love Story*, 61-62; Meikle, *American Plastic*, 180-193; Rogers, *Gone Tomorrow*, 122.

Mike explosion: Hargittai, *Judging Edward Teller*, 266; Rhodes, *The Making of the Atomic Bomb*, 777; Rhodes, *Dark Sun*, 500-510.

"Bomb pulse": Sam Scott, "What Bikini Atoll Looks Like Today: Sixty years after the nuclear tests, the groundwater is contaminated and the coconuts are radioactive. But are the coral reefs thriving?," *Stanford Magazine*, November 20, 2017.

III. THE LIVES

FREEPORT (189-193)

Dow's Freeport complex: "Industry," The Chamber Brazosport Area, https://www.brazosport.org/brazosport-is/industry/; Sharon Lerner, "The War on the War on Cancer," *The Intercept*, January 12, 2020; Chemical Retrieval on the Web (CROW), "Polymer Properties Database."

Magnesium: John Emsley, "Magnesium," The Royal Society of Chemistry, July 1, 2008, www.eic.rsc.org.

Graveyard of former enslaved people at Brazos Mall: St. John Barned-Smith, "In Lake Jackson, discovery of slave remains helps clarify town's history," *Houston Chronicle*, September 1, 2014; "Sugar in the dirt, excavations at Lake Jackson Plantation," Texas Beyond History, October 22, 2002, www.texasbeyondhistory.net.

Demographics of Freeport and Lake Jackson: "City Data Fact Sheets for Lake Jackson, Texas" DataUSA, accessed July 21, 2020, http://www.city-data.com/city/Lake-Jackson-Texas.html; "City Data Fact Sheets for Freeport, Texas," DataUSA, accessed July 21, 2020, http://www.city-data.com/city/Freeport-Texas.html; "LAKE JACKSON, TX," DataUSA, accessed June 2, 2020, www.datausa.io.

Air pollutants in Freeport: Environmental Protection Agency, "2017 Toxic Release Inventory Fact Sheet for Zip Code 77541," April 2019, www.enviro.epa.gov; Ari Phillips, "Plastics Pollution on the Rise," Environmental Integrity Project, September 5, 2019, 17-28, environmentalintegrity.org.

Asthma: "How Asthma Affects Your Body," American Lung Association, April 8, 2020, www.lung.org; "What Causes or Triggers Asthma," Asthma and Allergy Foundation of America, October 2019, www.aafa.org.

Health effects of ozone: Lee, Miller, and Shah, "Air Pollution and Stroke," 2-11; Ian Sample, "Air pollution now major contributor to stroke, global study finds," *The Guardian*, June 9, 2016; Catharine Paddock, "Ozone exposure tied to cardiovascular risk," *Medical News Today*, July 18, 2017; World Health Organization, "Top 10 Global Causes of Death, 2016," May 24, 2018, https://www.who.int/news-room/fact-sheets/detail/the-top-10-causes-of-death.

Texas Commission on Environmental Quality's lax permitting: Kiah Collier, "Environmental groups sue EPA over lax Texas air pollution permits," *Texas Tribune*, July 20, 2017; Gabriel Nelson, "By messing with Texas permits, EPA unleashes power struggle," *Environment and Energy News*, August 10, 2010.

Texas fights smog standards: Jamie Smith Hopkins, "Texas aligns itself with industry in fight against tighter smog standards," The Center for Public Integrity, March 18, 2015, www.publicintegrity.org; American Lung Association, "The State of the Air 2020," accessed June 2, 2020, www.stateoftheair.org/key-findings/.

LIGHTS (194-200)

Freeport's East End history: Melanie Oldham, Jessie Parker, and Manning Rollerson, interview with the author, July 5, 2019; Michael Barajas, "Freeport's East End Began in Segregation and Will End with Displacement," *Texas Observer,* September 12, 2018.

Dow's new plastic plant: Jordan Blum, "Dow Chemical completes crown jewel of $6B Gulf Coast expansion," Fuel Fix, March 28, 2017, www.fuelfix.com.

Plastic boom from fracking: Christopher M. Matthews, "The Shale Revolution's Staggering Impact in Just One Word, Plastics: Petrochemicals, once simply a cheap byproduct, are powering a U.S. manufacturing boom and export bonanza," *Wall Street Journal,* June 25, 2017; Marissa Luck, "Report details growing climate change risks from plastics industry," *Houston Chronicle*, May 15, 2019; "Plastics Pollution on the Rise," Environmental Integrity Project, September 5, 2019, 13-15, environmentalintegrity.org.

Oil and gas industry's pivot to plastic: "Plastic Packaging Market Size Worth $320.9 Billion By 2027," Grand View Research, April 2020, www.grandviewresearch.com.

Plastic industry promotes single-use plastic during pandemic: Joe Brock, "Pandemic Exposes Cracks in Oil Majors' Bet on Plastic," *Reuters,* June 4, 2020; Zoë Schlanger, "Will Coronavirus Be the Death or Salvation of Big Plastic?" *Time,* May 4, 2020; Sharon Lerner, "Big Plastic Asks For $1 Billion Coronavirus Bailout," *The Intercept*, April 27, 2020; Tony Radoszewski, "First Line of Defense with Products to Fight Coronavirus," Plastics Industry Association, March 20, 2020, www.plasticsindustry.org.

Environmental racism and plastic pollution: Mustafa Santiago Ali, "Environmental racism is killing Americans of color. Climate change will make it worse," *The Guardian*, July 28, 2020: www.theguardian.com; Zoë

Carpenter, "The Toxic Consequences of America's Plastics Boom," *The Nation*, March 14, 2019.

"I can't breathe": Mike Baker, Jennifer Valentino-DeVries, Manny Fernandez and Michael LaForgia, "Three Words. 70 Cases. The Tragic History of 'I Can't Breathe,'" *The New York Times,* June 29, 2020.

People of color in the U.S. at higher risk from coronavirus: Centers for Disease Control and Prevention, "Coronavirus Disease 2019 (COVID-19): People who are at higher risk for severe illness," May 14, 2020, www.cdc.gov; Centers for Disease Control and Prevention, "Coronavirus Disease 2019 (COVID-19): COVID-19 in Racial and Ethnic Minority Groups," April 22, 2020, www.cdc.gov; "Coronavirus in African Americans and Other People of Color. Featured expert: Sherita Hill Golden," Johns Hopkins Medicine, April 20, 2020, www.hopkinsmedicine.org; Mustafa Santiago Ali, "One reason why people of color are dying at higher rates in the US? The air they breathe," *The Guardian*, April 17, 2020.

RISK FACTOR (201-202)

Cancer clusters in Freeport: Texas Department of State Health Services, "Assessment of the Occurrence of Cancer"; "Liver Cancer: Statistics," American Society of Clinical Oncology, June 2019, www.cancer.net; Centers for Disease Control and Prevention, "Liver Cancer and the Environment," October 26, 2016, www.ephtracking.cdc.gov; Lerner, "The War on the War on Cancer"; Kathy Wong, "Emotions in Traditional Chinese Medicine," verywellmind, March 20, 2019, www.verywellmind.com.

Claudia Rankine and the liver: Rankine, *Don't Let Me Be Lonely*, 90.

HEARTLAND (203-207)

Heartland as a "psychic fallout shelter" for whites: Hoganson, *The Heartland*, xvi.

Port Freeport expansion: Environmental Protection Agency, "Ports Primer: 6.2 Port Factors Impacting the Regional Economy," last updated September 16, 2019, www.epa.gov; Ben Thrower, "Commentary: Expansion of the Panama Canal benefits global trade," Freightwaves, July 15, 2019, www.freightwaves.com; Dug Begley, "Port Freeport stakes its claim on cargo boom," *Houston Chronicle,* May 10, 2016.

They join generations of Black activists: Susan A. Mann, "Pioneers of U.S. Ecofeminism and Environmental Justice." Dorceta E. Taylor, "Race, Class, Gender, and American Environmentalism." See also: Robert D. Bullard,

Dumping in Dixie: Race, Class, and Environmental Quality," and Taylor's trilogy of books: *The Environment, and People in America's Cities: 1600s-1900s; Toxic Communities: Environmental Racism, Industrial Pollution and Residential Mobility*; and *The Rise of the American Conservation Movement: Power, Privilege, and Environmental Protection.*

East End resistance: Erin Callahan, "EAST END GAME: Freeport East Enders reluctant to relocate," *The Facts*, February 1, 2016; Michael Barajas, "Freeport's East End Began in Segregation and Will End with Displacement," *Texas Observer*, September 12, 2018; Lone Star Legal Aid, "Environmental Justice News," July 2019 Volume 2, Issue 7, https://myemail.constantcontact.com/Environmental-Justice-News---July-2019.html?soid=1113312890902&aid=o_AZ6s1LOJM; Elizabeth Parrish, "East End lawsuit against Port Freeport reaches federal court," *The Facts*, May 7, 2019 Amy Dinn (Lone Star Legal Aid), interview with author, June 21, 2019.

JOY (208-212)

"We have beauty in our lives": J Pluecker, conversation with the author, July 6, 2019.

***citysinging* exhibition:** Laura August, curator, June 21–August 18, 2019, exhibition catalogue: https://lawndaleartcenter.org.

ITC chemical fire: Deer Park, Texas, Local Emergency Planning Committee, undated, http://www.deerparktx.gov/1722/Shelter-In-Place; Julian Gill, "Deer Park plant fire: What you need to know about benzene," *Houston Chronicle*, March 18, 2019; Perla Trevizo, "The ITC fire created 20 million gallons of waste. Getting rid of it is no easy task." *Houston Chronicle*, July 21, 2019.

Doomsday clock: Bulletin of the Atomic Scientists, "Doomsday Clock," https://thebulletin.org/doomsday-clock/faq/. In 2020, they set the clock at 90 minutes to midnight.

"It's a way of holding your hand": Julia Landois Barbosa, interview with the author, August 24, 2020.

***Leroy's* and the joy of Blues clubs:** Robert Hodge, interview with the author, September 2, 2020.

REMEMBER (213-217)

East End Freeport lawsuit dismissed: Rollerson v. Port Freeport, 3:2018-cv-00235 (txsd); Amy Dinn, email correspondence with the author, October 2019.

Port Freeport eminent domain: Maddy McCarty, "Port Freeport proceeds with eminent domain of East End," *The Facts*, October 11, 2019.

Brazos Mall and the gravesites of former enslaved people: Sara Benner, "Unmarked graves discovered adjacent to slave cemetery," *The Facts*, August 22, 2014; Andy Packard, "Brazos Mall dedicates historical Mount Zion cemetery," *The Facts*, September 13, 2015.

JOY (218-221)

Freeport LNG and Quintana, Texas: U.S. Energy Information Administration, "Natural Gas Explained: Liquefied Natural Gas," June 4, 2019, https://www.eia.gov/energyexplained/index.php?page=natural_gas_lng; Naveena Sadasivam, "Company Town," *Texas Observer*, May 31, 2016; Ellen M. Gilmer, "Environmentalists lost big on LNG exports. Now what?" *Energywire*, February 21, 2018; Ron Bousso, "LNG growth to propel oil and gas industry's carbon emissions," *Reuters*, September 29, 2017; Umair Irfan, "Report: we have just 12 years to limit devastating global warming," *Vox*, October 8, 2018; United Nations Intergovernmental Panel on Climate Change, "Summary for Policymakers of IPCC Special Report on Global Warming of 1.5°C approved by governments," October 8, 2018; Melanie Oldham, interview with the author, August 2019.

"Even a wounded world is feeding us": Robin Wall Kimmerer, *Braiding Sweetgrass* (Minneapolis, MN: Milkweed, 2013), 327.

"There is a joy to the struggle": J Pluecker, email correspondence with the author, September 2020.

"I'm not giving up": Melanie Oldham, interview with the author, July 5, 2019.

"I'm not afraid": Manning Rollerson, interview with the author, July 5, 2019.

DIS-COURAGE (222-227)

"Oh it's just that Summerlin guy again": Errol Summerlin, tour with the author of Exxon-SABIC plant and Corpus Christi, July 4, 2020.

Exxon-SABIC plant: Jordan Blum, "$10 billion Exxon-Saudi project concerns Texas communities," *Houston Chronicle*, January 29, 2017; Martin Chulov, "Saudi leaders hail Trump visit as 'reset of regional order,'" *The Guardian*, May 20, 2017; Oliver Milman, "World's largest plastics plant rings alarm bells on Texas coast," *The Guardian*, December 26, 2017; Renée Feltz, "Plastics on Hurricane Alley," *The Progressive*, December 1, 2018; Zoë Carpenter, "The Toxic Consequences of America's Plastics Boom," *The Nation*, March 14, 2009.

Industry's use of codenames for new projects: Jon Chesto, "Code names become the norm for business relocation efforts," *Boston Globe*, January

26, 2016; Tamara Chuang, "Project code name: How firms seeking state aid keep names private," *Denver Post*, July 29, 2015.

Texas Commission on Environmental Quality's lax permitting: Kiah Collier, "Environmental groups sue EPA over lax Texas air pollution permits," *Texas Tribune*, July 20, 2017, www.texastribune.org; Gabriel Nelson, "By messing with Texas permits, EPA unleashes power struggle," *Environment and Energy News*, August 10, 2010, www.eenews.net.

Fracking in the Permian Basin: Ariel Cohen, "America's Oil And Gas Reserves Double With Massive New Permian Discovery," *Forbes*, December 21, 2018; Clifford Krauss, "The 'Monster' Texas Oil Field That Made the U.S. a Star in the World Market," *The New York Times*, February 3, 2019, www.nytimes.com; Olivia Judson, "When Texas Was at the Bottom of the Sea," *Smithsonian Magazine*, January 2015; Natural Gas Intelligence, "Information on the Permian Basin," *Shale Daily*, accessed August 26, 2019, www.naturalgasintel.com.

"**Stay alive on 285":** Jordan Blum, "Texas' most dangerous border leads to New Mexico," *Houston Chronicle*, October 3, 2019; "'Death Highway' in Texas' Permian Basin sees accidents, fatalities pick up as oil price rises," *Dallas News*, July 30, 2018; Renee Lewis, "How Texas' oil fracking boom tore a 'highway of death' through this tiny town," *Vice*, November 12, 2018.

The sense of threat in the room was visceral: Coastal Alliance to Protect our Environment (CAPE) meeting, Corpus Christi, July 3, 2019; Sergio Chapa, "EPIC delivers first Permian Basin crude oil to Port of Corpus Christi," *Houston Chronicle*, August 29, 2019; Sergio Chapa, "Report: Port of Corpus Christi to become top U.S. crude oil export hub," *Houston Chronicle*, June 11, 2019.

The kind of project only vast wealth could build: Gulf Coast Growth Ventures, "GCGV Project Animation," accessed 25 September 2019, www.gulfcoastgv.com/videos; Gulf Coast Growth Ventures, "Gulf Coast Growth Ventures Project Update, September 2019," accessed 25 September 2019, www.gulfcoastgv.com/videos.

WE'LL BE HERE (228-230)

Diane Wilson victory against Formosa: Michael Berryhill, "The Battle for Lavaca Bay," *Houston Press*, May 19, 1994; Lily Moore-Eissenberg, "Nurdles All the Way Down," *Texas Monthly*, September 18, 2019; Carlos Anchondo, "Environmentalists take petrochemical giant Formosa to court over plastics pollution," *Texas Tribune*, March 25, 2019; Amal Ahmed, "Nurdle by Nurdle, Citizens Took on A Billion-Dollar Plastic Company — and Won," *Texas*

Observer, July 3, 2019; Kiah Collier, "Retired Texas shrimper wins record-breaking $50 million settlement from plastics manufacturing giant," *Texas Tribune*, December 3, 2019.

"We'll be here": Diane Wilson, quoted in Ahmed, "Nurdle by Nurdle."

New Formosa plant in Louisiana: Center for Biological Diversity, "State of Louisiana Approves Key Permit for Toxic Polluter in Cancer Alley," February 1, 2019, www.biologicaldiversity.org; Kristen Hays, "FG LA planning construction soon on $9.4 billion Louisiana petrochemical complex," *S&P Global*, February 4, 2020, www.spglobal.com.

HERE (231-234)

Corpus Christi desalination plants: Tim Acosta, "Corpus Christi desalination plans could soon become less murky," *Corpus Christi Caller Times*, June 12, 2019; Greg Chandler, "Coastal Bend desalination plant closer to reality," *KRISTV*, August 28, 2019, www.kristv.com.

Former Exxon executive on Texas Water Development Board: Dan Wallach, "ExxonMobil spokeswoman, LNVA board member Kathleen Jackson moves to Texas Water Development Board as one of three commissioners," *Beaumont Enterprise*, March 14, 2014; Chris Ramirez, "Corpus Christi applies for $222 million loan for desalination plant in port's Inner Harbor," *Corpus Christi Caller Times*, April 23, 2020.

Desalination's harmful impacts: Alister Doyle, "Too much salt: water desalination plants harm environment: U.N.," *Reuters*, January 14, 2019; Tik Root, "Desalination plants produce more waste brine than thought," *National Geographic*, January 14, 2019; Texas Campaign for the Environment, "TCE and For the Greater Good Scale Up NO DESAL City Charter Amendment Petition with Drive-By Signings," June 18, 2020.

School desegregation in Corpus Christi: Araiza and Marquez, "Cisneros v. CCISD"; Dr. Isabel Araiza, interview with the author, July 23, 2019.

Industrial contamination in Corpus Christi's water supply: Allison Ehrlich, "Timeline: Corpus Christi water issues since 2015," *Corpus Christi Caller Times*, December 15, 2016; Kiah Collier, "In Corpus Christi, the latest water emergency was different," *Texas Tribune*, December 21, 2016; Araiza, interview with the author, July 23, 2019.

For the Greater Good national news coverage: "Corpus Christi water ban: Resident calls it a 'huge failure' of city," *CBS News*, December 16, 2016.

The "downward spiral" of industrial tax breaks: Hochschild, *Strangers in the Their Own Land*, 73-78; Araiza, interview with the author, July 23, 2019.

TIRED (235-238)

Legacy of pollution in Mossville, Louisiana: Helfand and Gold, *Blue Vinyl*; Frankland, *Women Pioneers*, 80; Wilma Subra, interview with the author, June 20, 2019.

Activists get tired: Heather Rogers, "Erasing Mossville," *The Intercept*, November 4, 2015, www.theintercept.com.

History of SASOL in Apartheid South Africa and the company's environmental impact: Daniel Gross, "Thanks for the Cheap Gas, Mr. Hitler!" *Slate*, October 23, 2006; African Climate Reality Project, "Climate Change in South Africa: Who are the Carbon Criminals?" accessed June 13, 2020, www.climatereality.co; Hallowes, *How Corporations Rule*.

Mossville's history and struggles with pollution, SASOL buying out the town to expand plastic plant: Rick Mullin, "Mossville's end: As Sasol's huge petrochemical project lifts Southwest Louisiana, an environmental justice community dissolves in its shadow," *Chemical and Engineering News*, March 21, 2016, www.cen.acs.org; "Sasol says U.S. ethane cracker production steps up," *Reuters*, August 28, 2019; Tedd Griggs, "After Sasol ditches plan for $11B plant, Louisiana off hook for $200M in incentives," *Baton Rouge Advocate*, November 27, 2017; Tim Murphy, "A Massive Chemical Plant Is Poised to Wipe This Louisiana Town off the Map," *Mother Jones*, March 27, 2014; Katherine Sayre, "Closing Costs: As a chemical plant expands, Mossville, Louisiana, vanishes," *New Orleans Times-Picayune*, November 15, 2017; Heather Rogers, "Erasing Mossville," *The Intercept*, November 4, 2015; David S. Martin, "Toxic towns: People of Mossville 'are like an experiment'," *CNN*, February 26, 2010; Wilma Subra, interview with the author, June 20, 2019; see also the documentary "Mossville: When Great Trees Fall" by Alex Glustrom, 2019, http://www.mossvilleproject.com/

"That tired old Mossville story": Rick Mullin, "Mossville's end," *Chemical and Engineering News*, March 21, 2016.

"We're talked out. We're tired.": Christine Bennett, phone conversation with the author, July 4, 2019.

LIFE (239-242)

On apologies: Oliner, *Altruism*, 65; Lazare, *On Apology*, 229-263.

Mossville oral histories: "Mossville History Project," T. Harry Williams Center for Oral History, Louisiana State University Libraries and Imperial Calcasieu Museum, www.lib.lsu.edu/oralhistory.

"Somebody was always throwing a party": Bennett and Bennett, "Delma and Christine Bennett Interview Session I," transcript 60.

"There was nothing for us to be afraid of": Ambrose, "Lenoria Braxton Ambrose Interview Session II," transcript 20-21.

"How can you pay somebody for life?": Bennett and Bennett, "Delma and Christine Bennett Interview," transcript 11, 24.

BLANKS (243-247)

Atakapa Ishak history: Atakapa Ishak Nation, "History," accessed June 22, 2020, www.atakapa-ishak.org; Wikipedia, "Atakapa," accessed June 22, 2020, https://en.wikipedia.org/wiki/Atakapa; Eric Besson, "SE Texas' Atakapa tribe seeking federal designation," *Beaumont Enterprise*, September 12, 2014.

History of Calcasieu Parish: Ellender, "A Brief History of Calcasieu Parish."

Charles Sallier: City of Lake Charles, "History of Lake Charles," accessed June 22, 2020, www.cityoflakecharles.com.

Nell Chennault: "Life Goes Calling on Mrs. Chennault: General's wife spends a busy Sunday at home in Water Proof, La.," *Life*, March 15, 1943, 98-101.

Anna Chennault: Matt Schudel, "Anna Chennault, secret Nixon envoy and Washington figure of 'glamour and mystery,' dies at 94," *The Washington Post*, April 3, 2018; John A. Farrell, "Anna Chennault: The Secret Go-Between Who Helped Tip the 1968 Election," *Politico*, December 30, 2018; Catherine Forslund, *Anna Chennault: Informal Diplomacy and Asian Relations* (Lanham, Maryland: Rowman & Littlefield, 2002).

SILENCE (248-252)

The Mossville school: Felix and Shelton, "Dorothy Felix and Evelyn Gasaway Shelton Interview Session I," transcript 19.

Violence of school integration in Mossville: Jones, "Brenda Cole Jones Interview Session I," transcript 43; Latour et al., "Jourdan Family Interview Session I," transcript 110.

"Do I say just really what happened?": Bennett and Bennett, "Delma and Christine Bennett Interview," transcript 4-5.

Mossville residents' experience of plants and pollution: Dotson, "Daren Doston Sr. Interview Session I," transcript 17-18; Bennett and Bennett, "Delma and Christine Bennett Interview," transcript; Hunstberry, "Jawanna Huntsberry Interview Session I," transcript 75-76; Jones, "Brenda Cole Jones Interview Session I," transcript 68; Frankland, *Women Pioneers*, 81-82.

Dorothy Hartman Felix and the human rights petition from Mossville residents: Wilma Subra, interview with the author, June 20, 2019; Sue Sturgis, "Louisiana environmental racism case gets hearing from Inter-American Commission on Human Rights," *Grist*, April 2, 2010; Felix and Shelton, "Dorothy Felix and Evelyn Gasaway Shelton Interview Session I," transcript 33-35.

ALIVE (253-257)

Morning Star Cemetery: Dotson, "Della Dotson Interview Session I," transcript 23.

Christine Bennett and others want more than a museum exhibit and a set of oral histories: "History of Mossville exhibit opens," *KLPCTV News,* June 9, 2017; Christine Bennett, conversation with the author, July 5, 2019.

Historically Black communities across Louisiana wiped out by industrial expansion: Rebecca Johnson, "The Exodus of the People of Mossville," *Reimagine Race, Poverty and The Environment*, June 5, 2017, www.reimaginerpe.org; David E. Newton, "Environmental Justice: a Reference Handbook," (Santa Barbara, California: ABL-CIO, 2009), 10; Luna Reyna, "Environmental Racism Is Killing Black Communities In Louisiana," *Talk Poverty*, January 9, 2020, www.talkpoverty.org; Rachel Ramirez, "These Louisiana activists are facing 'terrorizing' charges for a stunt they pulled 6 months ago," *Grist*, June 25, 2020.

"Nothing can be allowed to escape": Robin McDowell, "Black Resistance in Louisiana's Cancer Alley," *Boston Review*, June 4, 2019.

"What has been dead is now alive": Lorde, "The Poet as Outsider," *Dream of Europe*, 103. Lorde uses this phrase in response to "Poem at Thirty" by Sonia Sanchez, which ends with these lines:

> you you black man
> stretching scraping
> the mold from your body.
> here is my hand.
> i am not afraid
> of the night.

TRUST (258-261)

Nuclear weapons testing in the Marshall Islands: Walsh et al., *Etto n̄an Raan Kein*, 291; Parsons and Zaballa, *Bombing the Marshal Islands*, 16-20.

"The world's first nuclear disaster": Parsons and Zaballa, *Bombing the Marshall Islands*, 26.

U.S. view of the Marshalls as a testing ground: Walsh et al., *Etto n̄an Raan Kein*, 283-285; Smith-Norris, *Domination and Resistance*, 4-6.

"He just looked through them": Parsons and Zaballa, *Bombing the Marshall Islands*, 127.

The U.S. exploded all of its largest thermonuclear weapons in the Marshalls: Parsons and Zaballa, *Bombing the Marshall Islands*, 2; Smith-Norris, *Domination and Resistance*, 5-6; Walsh et al., *Etto n̄an Raan Kein*, 310-11; Ali Raj, "In Marshall Islands, radiation threatens tradition of handing down stories by song," *Los Angeles Times*, November 10, 2019.

Legacy of contamination on Bikini and Enewetak atolls: Walsh et al., *Etto n̄an Raan Kein*, 296-305; Parsons and Zaballa, *Bombing the Marshall Islands*, 52; Bikini Atoll "Basic Facts," accessed 30 December 2019, www.bikiniatoll.com; Barad, "After the End of the World," 536, uses the term "nuclear refugee" and it also appears in other accounts of the Marshall Islands; Susanne Rust, "Radiation in parts of the Marshall Islands is far higher than Chernobyl, study says," *Los Angeles Times*, July 15, 2019; Ariana Rowberry, "Castle Bravo: The Largest U.S. Nuclear Explosion," *Brookings*, February 27, 2014.

HOLE (262-265)

ri-Enewetak in World War II: Walsh et al., *Etto n̄an Raan Kein*, 246-280; "U.S. Troops Capture the Marshall Islands," *History*, July 28, 2019, www.history.com.

***Mej*, the word for death in Marshallese:** *Marshallese-English Online Dictionary*, updated August 31, 2019, www.trussel2.com/mod/MED2M.htm#mej.

"You are like a rabbit fish wriggling on the end of a spear": Smith-Norris, *Domination and Resistance*, 17.

The U.S. forbid the ri-Enewetak from describing the blast: Smith-Norris, *Domination and Resistance*, 17-18.

Footage and oral testimony of Mike explosion: Joint Task Force 132, "Operation Ivy," 1952, https://archive.org/details/OperationIVY1952; Rhodes, *Dark Sun*, 508-510.

BOY (266-267)

Ulam justifies his work on the thermonuclear bomb: Ulam, *Adventures of a Mathematician*, 221-22.

"It's a boy": Rhodes, *Dark Sun*, 510-511.

CIRCLE (268-270)

"Basket": Jetñil-Kijiner, *Iep Jāltok: Poems from a Marshallese Daughter*, 4-5.

"He put all his children, grandchildren, and wife in a hole": Anonymous, "My Great Great Grandfather," Lolelaplap Voices, April 29, 2013, www.marshallislandsvoices.blogspot.com.

The continued generations that hole made possible: UN Climate Change, "COP24 Live Conversation with Kathy Jetñil-Kijiner."

History of the ouroboros: "Ouroboros," Britannica, accessed July 1, 2020, www.britannica.com; Joobin Bekhrad, "The ancient symbol that spanned millennia," *BBC*, December 4, 2017.

"How shall we remember you?": This is a line from another poem by Jetñil-Kijiner, "Anointed," cited by Barad. This whole discussion is indebted to the work of both Jetñil-Kijiner and Barad. The video production of this poem is available at: https://www.kathyjetnilkijiner.com/videos-featuring-kathy.

PULSE (271-274)

Radiation health effects amongst Marshallese: Smith-Norris, *Domination and Resistance*, 6-7; Walsh et al., *Etto n̄an Raan Kein*, 299-302; Parsons and Zaballa, *Bombing the Marshall Islands*, 81-82.

"I get invited to more funerals than I do birthday parties": Kianna Juda-Angelo, interview with the author, February 3, 2020.

Marshallese in the U.S. suffer disproportionately from COVID-19: Matsumoto, Frost, and Cortez, "Pacific Islanders Disproportionately Affected by COVID-19"; Alex Golden and Doug Thompson, "Marshallese contracting, dying from COVID-19 at disproportionate rate," *Arkansas Democrat Gazette*, June 14, 2020; Zuzanna Sitek, "Northwest Arkansas Struggles To Deal With A Surge In Coronavirus Cases," *All Things Considered*, National Public Radio, July 1, 2020; Deepa Bharath, "How the coronavirus is devastating Southern California's Pacific Islanders," *Orange County Register*, June 29, 2020; Jim Mendoza, "Travel ban leaves 100s of Marshallese stranded in Hawaii, waiting for the all-clear to go home," *Hawaii News Now*, June 18, 2020.

THE TOMB (275-277)

ri-Enewetak struggle to return home and ongoing contamination: Smith-Norris, *Domination and Resistance*, 10, 17; Parsons and Zaballa, *Bomb-*

ing the Marshall Islands, 157; Susanne Rust, "How the U.S. betrayed the Marshall Islands, kindling the next nuclear disaster," *Los Angeles Times*, November 10, 2019.

Marshall Islands constitution: Walsh et al., *Etto n̄an Raan Kein*, 379.

Construction of the Tomb: Parsons and Zaballa, *Domination and Resistance*, 16-43; Terry Hamilton, "Report"; Susanne Rust, "15 months, 5 trips, a gut-wrenching sight: How we reported the Marshall Islands story," *Los Angeles Times*, November 10, 2019.

BREATHING (275-281)

"It looked like it was breathing": Susanne Rust and Carolyn Cole, "High radiation levels found in giant clams near U.S. nuclear dump in Marshall Islands," *Los Angeles Times*, May 28, 2019.

Ongoing threats posed by the Tomb: Evan Lubovsky, "Putting the 'nuclear coffin' in perspective," Woods Hole Oceanographic Institution, August 13, 2019, www.whoi.edu; Susanne Rust, "New U.S. law requires probe of Marshall Islands nuclear dump threatened by rising seas," *Los Angeles Times*, December 30, 2019; Rust and Cole, "High radiation levels found in giant clams," *Lost Angeles Times*; Hamilton, "Report."

Jo-Jikum: "Jo-Jikum," accessed July 7, 2020, www.jojikum.org.

Struggle to manage waste in the Marshall Islands: "Einwot Juon," Facebook Profile, accessed July 7, 2020, https://www.facebook.com/EinwotJuon/?eid=ARAzFpXcwx_jIG8di_a4gaWiAJUNnxaw8IHa7msuCe-EX_aIUfEB-p-KvnvP_rssFUPoFFLoFP18TubuS; "Incinerators shipped to Marshall Islands," *Solid Waste & Recycling*, January 15, 2019; Asian Development Bank, "Solid Waste Management in the Pacific: The Marshall Islands Country Snapshot," June 2014, https://www.adb.org/sites/default/files/publication/42669/solid-waste management-marshall-islands.pdf.

"We cannot think about abandoning the land": Suzanne Rust and Carolyn Cole, "How the U.S. betrayed the Marshall Islands, kindling the next nuclear disaster," *Los Angeles Times*, November 10, 2019.

"There's a grief": UN Climate Change, "COP24 Live Conversation with Kathy Jetn̄il-Kijiner."

BURN (282-287)

Health effects of benzene: American Cancer Society, "What Is Acute Myeloid Leukemia (AML)?" August 21, 2018, www.cancer.org; Leukemia & Lymphoma Society, "Acute myeloid leukemia (AML)," accessed February

15, 2020, www.lls.org; National Cancer Institute, "Adult Acute Myeloid Leukemia Treatment (PDQ®)–Health Professional Version," January 22, 2020, accessed February 15, 2020, www.cancer.gov; World Health Organization, "Exposure to Benzene: A Major Public Health Concern," 2010, who.int/ipcs/features/benzene.pdf.

Texas has the highest rate of benzene pollution in the U.S.: Allyn West, "What is Benzene," One Breath Partnership, November 1, 2019; Christopher Collins, "Six Texas Oil Refineries Are Among the Nation's Worst Benzene Polluters, Data Shows," *Texas Observer*, February 6, 2020.

Health effects of plastic pollution: Azoulay et al., "Plastic & Health" 17-27; Science Direct, "Aromatic Hydrocarbon," 2020.

Climate damage from plastic pollution: Hamilton and Felt, *Plastic & Climate*.

Damage vs desire-centered inquiry: Tuck, "Suspending Damage, a Letter to Communities." I am grateful to Irene Vázquez for pointing me to this reference.

"Who anointed *them* with the power to burn?": Kathy Jetñil-Kijiner, "Anointed," April 16, 2018, www.kathyjetnilkijiner.com.

ACKNOWLEDGMENTS (338-340)

Inexpert investigation: Boykoff and Sand, "Notes on inexpert investigation;" Joan Retallack, *The Poethical Wager*. I am indebted in particular to Kaia Sand for the idea of the inexpert in our conversations over the years.

SELECTED BIBLIOGRAPHY

Ahmed, Sara. *Queer Phenomenology: Orientations, Objects, Others.* Durham and London: Duke University Press, 2006.

Allen, J.A. *Studies in Innovation in the Steel and Chemical Industries*. New York: Augustus M. Kelley, 1968.

Ambrose, Lenoria Braxton. "Lenoria Braxton Ambrose Interview Session II." Interviewed by Chelsea Arsenault. Mossville History Project, LSU Libraries Special Collections Williams Center for Oral History (June 3, 2015). https://lib.lsu.edu/sites/all/files/oralhistory/presentations/mossvilledirectory/transcript/Ambrose_Lenoria_T4484.pdf.

American Chemical Society National Historic Chemical Landmarks. "Leo Hendrick Baekeland and the Invention of Bakelite." November 9, 1993. http://www.acs.org/content/acs/en/education/whatischemistry/landmarks/bakelite.html.

Araiza, Dr. Isabel, and Amanda Marquez. "Cisneros v. CCISD: A Community Divided." Del Mar College Lecture. Recorded March 2, 2016. YouTube video, 1:10:56. https://www.youtube.com/watch?v=EK5J2-D1kGk.

Arnold, James R. *The Moro War: How America Battled a Muslim Insurgency in the Philippine Jungle, 1902-1913*. New York: Bloomsbury Press, 2011.

Azoulay, David, Priscilla Villa, Yvette Arellano, Miriam Gordon, Doun Moon, Kathryn Miller, and Kristen Thompson. "Plastic & Health, the Hidden Costs of a Plastic Planet." Edited by Amanda Kistler. Center for International Environmental Law, February 2019. https://www.ciel.org/wp-content/uploads/2019/02/Plastic-and-Health-The-Hidden-Costs-of-a-Plastic-Planet-February-2019.pdf.

Barad, Karen. "After the End of the World: Entangled Nuclear Colonialisms, Matters of Force, and the Material Force of Justice." *Theory & Event* 22, no. 3 (2019): 524-550. muse.jhu.edu/article/729449.

——. *Meeting the Universe Halfway: Quantum Physics and the Entanglement of Matter and Meaning*. Durham, NC: Duke University Press, 2007.

——. "Quantum Entanglements and Hauntological Relations of Inheritance: Dis/continuities, SpaceTime Enfoldings, and Justice-to-Come," *Derrida Today* 3, no. 2 (2010): 240–268. DOI: 10.3366/E1754850010000813.

Barajas, Michael. "Freeport's East End Began in Segregation and Will End with Displacement." *Texas Observer*, September 12, 2018. www.texasobserver.org.

Barnes, David K.A., Francois Galgani, Richard C. Thompson and Morton Barlaz. "Accumulation and Fragmentation of Plastic Debris in Global Environments." *Philosophical Transactions. Biological Sciences of the Royal Society* 364, no. 1526 (July 2009): 1985-1998. DOI: 10.1098/rstb.2008.0205.

Bengali, Shashank. "Your Trash Is Suffocating this Indonesian Village. Here's How." *Los Angeles Times,* October 25, 2019. https://www.latimes.com/world-nation/story/2019-10-25/plastic-pollution-waste-recycling-indonesia.

Bennett, Delma, and Christine Bennett. "Delma and Christine Bennett Interview Session I." Interviewed by Chelsea Arsenault. Mossville History Project, LSU Libraries Special Collections Williams Center for Oral History (May 12, 2016): transcript 60. https://lib.lsu.edu/sites/all/files/oralhistory/presentations/mossvilledirectory/transcript/Bennett_Delma_and_Christine_T4620.pdf.

Bennett, Jane. *Vibrant Matter: A Political Ecology of Things.* Durham and London: Duke University Press, 2010.

Berkeley Lab. "Atom by Atom, Bond by Bond, a Chemical Reaction Caught in the Act." *Berkeley Lab* [News release], May 30, 2013. https://newscenter.lbl.gov/2013/05/30/atom-by-atom/.

Binus, Joshua. "Russian Old Believers." *The Oregon History Project,* last updated March 17, 2018. https://oregonhistoryproject.org/articles/historical-records/russian-old-believers/#.X0a-m9NKhAY.

Boykoff, Jules and Kaia Sand. "Notes on inexpert investigation." *Jacket* 2, June 19, 2011. jacket2.org.

Brahney, Janice, Margaret Hallerud, Eric Heim, Maura Hahnenberger, and Suja Sukumaran. "Plastic Rain in Protected Areas of the United States." *Science (American Association for the Advancement of Science)* 368. no. 6496 (2020): 1257-1260. DOI: 10.1126/science.aaz5819

Braxton, George, and Velma Carheel. "George Braxton and Velma Carheel Interview Session I." Interviewed by Stephanie Dragoon. Mossville History Project, LSU Libraries Special Collections Williams Center for Oral History (July 27, 2015): transcript 33. https://lib.lsu.edu/sites/all/files/oralhistory/presentations/mossvilledirectory/transcript/Braxton_George_and_Carheel_Velma_T4489.pdf.

Brown, Cecil. *Suez to Singapore*. New York: Random House, 1942.

Brown, Louis. *A Radar History of World War II: Technical and Military Imperatives.* Bristol and Philadelphia: Institute of Physics Publishing, 1999.

Browne, Malcolm W. "A Pervasive Molecule Is Captured in a Photograph." *The New York Times*, August 16, 1988. https://www.nytimes.com/1988/08/16/science/a-pervasive-molecule-is-captured-in-a-photograph.html.

Browne, Mark Anthony, Stewart J. Niven, Tamara S. Galloway, Steve J. Rowland and Richard C. Thompson. "Microplastic Moves Pollutants and Additives to Worms, Reducing Functions Linked to Health and Biodiversity." *Current Biology* 23, no. 23 (December 2, 2013): 2388-2392. DOI: 10.1016/j.cub.2013.10.012.

Brunello, Franco. *The Art of Dyeing in the History of Mankind*. Translated by Bernard Hickey. Venice: Neri Pozza, 1973.

Buderi, Robert. *The Invention that Changed the World: How a Small Group of Radar Pioneers Won the Second World War and Launched a Technological Revolution*. New York: Simon and Schuster, 1996.

Bullard, Robert D. *Dumping in Dixie: Race, Class, and Environmental Quality*, 3rd edition. Boulder CO: Westview Press, 2000.

Chemical Retrieval on the Web (CROW). "Polymer Properties Database." Copyright 2015-2020. www.polymerdatabase.com.

Christman, Elwyn. "Action Report of Commanding Officer, Patrol Plane 101-P-9 to Commander Patrol Wing 10, Subject: Bombing attack at Jolo, Sulu, December 27, 1941." College Park, MD: National Archives and Records Administration.

Coleridge, Samuel Taylor. "The Rime of The Ancient Mariner (Text of 1834)." https://www.poetryfoundation.org/poems/43997/the-rime-of-the-ancient-mariner-text-of-1834.

Colorants Industry History. "William H. Perkin: Founder of Dystuff Industry." Copyright 2016. http://www.colorantshistory.org/PerkinBiography.html.

Cooper, Necia Grant, Roger Eckhardt, and Nancy Shera, eds. *From Cardinals to Chaos: Reflections on the Life of Stanislaw Ulam*. Cambridge: Cambridge University Press, 1989.

Cox, Kieran D, Garth A. Covernton, Hailey L. Davies, John F. Dower, Francis Juanes, and Sarah E. Dudas. "Human Consumption of Microplastics." *Environmental Science and Technology* 53, no. 12 (2019): 7068-7074. DOI: 10.1021/acs.est.9b01517.

Creed, Roscoe. *PBY: The Catalina Flying Boat*. Annapolis: Naval Institute Press, 1985.

Crocker, Mel. *Black Cats and Dumbos: WWII's Fighting PBYs.* Blue Ridge Summit, PA: Aero Publishers, 1987.

D'Angina, James. *Mitsubishi A6M Zero*. New York: Osprey Publishing, 2016.

Dauxois, Thierry. "Fermi, Pasta, Ulam, and a Mysterious Lady." *Physics Today* 61, no. 1 (2008): 55-57. http://perso.ens-lyon.fr/thierry.dauxois/PAPERS/pt61_55.2008.pdf.

Dawley, J.B. "Bombing attack at Jolo, Sulu, December 27, 1941 and subsequent incidents, Commanding Officer, Patrol Plane 101-P-6 to Commander, Patrol Wing 10, January 29, 1942." College Park, MD: National Archives and Records Administration.

Dolan, Therese. "The Empress's New Clothes: Fashion and Politics in Second Empire France." *Woman's Art Journal* 15, no. 1 (1994): 22-28. DOI: 10.2307/1358491.

Dorny, Louis B. *US Navy PBY Catalina Units of the Pacific War*. Botley, Oxford: Osprey Publishing, 2007.

Dotson, Daren Sr. "Daren Dotson Sr. Interview Session I," Interviewed by Douglas Mungin. Mossville History Project, LSU Libraries Special Collections Williams Center for Oral History (January 24, 2015): transcript 17-18. https://lib.lsu.edu/sites/all/files/oralhistory/presentations/mossville directory/transcript/Dotson_Daren_Sr_T4425.pdf.

Dotson, Della. "Della Dotson Interview Session I." Interviewed by Rebecca Cooper. Mossville History Project, LSU Libraries Special Collections Williams Center for Oral History (July 27, 2015): transcript 12. https://lib.lsu.edu/sites/all/files/oralhistory/presentations/mossvilledirectory/transcript/Dotson_Della_T4488.pdf.

Ebbesmeyer, Curtis and Eric Scigliano. *Flotsametrics and the Floating World: How One Man's Obsession with Runaway Sneakers and Rubber Ducks Revolutionized Ocean Science.* New York: HarperCollins, 2010.

Ellender, Allie. "A Brief History of Calcasieu Parish." Transcribed by Leora White. McNeese State University Library Archives (May 2007). http://ereserves.mcneese.edu/depts/archive/FTBooks/ellender%20 history.htm.

Felix, Dorothy, and Evelyn Gasaway Shelton. "Dorothy Felix and Evelyn Gasaway Shelton Interview Session I." Interviewed by Jennifer A. Cramer. Mossville History Project, LSU Libraries Special Collections Williams Center for Oral History (January 23, 2015): transcript 19. https://lib.lsu.edu/sites/all/files/oralhistory/presentations/mossvilledirectory/transcript/Felix_Dorothy_and_Gasaway_Evelyn_Shelton_T4404.pdf.

Fenichell, Stephen. *Plastic: The Making of a Synthetic Century*. New York: HarperBusiness, 1997.

Fermi, Enrico. "Possible Production of Elements of Atomic Number Higher than 92." *Nature (London)* 133, no. 3372 (1934): 898-899. DOI:10.1038/133898a0.

Filarowski, Andrew. "Perkin's Mauve: The History of the Chemistry." *Resonance* 15, no. 9 (2010): 850-855. DOI: 10.1007/s12045-010-0094-5.

Forslund, Catherine. *Anna Chennault: Informal Diplomacy and Asian Relations*. Lanham, Maryland: Rowman & Littlefield, 2002.

Frankland, Peggy, with Susan Tucker. *Women Pioneers of the Louisiana Environmental Movement*. Jackson, MI: University of Mississippi Press, 2013.

Freinkel, Susan. *Plastic: A Toxic Love Story*. Boston, Massachusetts: Houghton Mifflin Harcourt, 2011.

Fritz, W. Barkley. "The Women of ENIAC." *IEEE Annals of the History of Computing* 18, no. 3 (1996): 13-28. DOI: 10.1109/85.511940.

Garcia, Jerry. "Latinos in Oregon." *The Oregon Encyclopedia*, last updated September 17, 2019. https://oregonencyclopedia.org/articles/hispanics_in_oregon/#.Xoa1-tNKhAY.

Garfield, Simon. *Mauve: How One Man Invented a Color That Changed the World*. New York and London: W.W. Norton and Co., 2001.

General Electric. *How Plastics Solved War Problems, 24 Case Histories*. Pittsfield, MA: General Electric Chemical Department, Plastics Division, 1946.

Geyer, Roland, Jenna R. Jambeck, and Kara Lavender Law. "Production, Use, and Fate of All Plastics Ever Made." *Science Advances* 3, no. 7 (2017): E1700782. https://advances.sciencemag.org/content/3/7/e1700782.full.

Glustrom, Alex. *Mossville: When Great Trees Fall*. 2019. http://www.mossvilleproject.com/.

Goldman, Judith. "The Dispossessions [; or, When I got back to the changing shed, the Albatross [." In *l.b.; or catenaries*. San Francisco: Krupskaya, 2011, 139.

Goodsell, David. "Fact and Fantasy in Nanotech Imagery." *Leonardo (Oxford)* 42, no.1 (2009): 52-57. DOI: 10.1162/leon.2009.42.1.52.

Habashi, Fathi. "Ida Noddack: Proposer of nuclear fission." In *A Devotion to Their Science: Pioneer Women of Radioactivity*, edited by Marlene F. Rayner-Canham and Geoffrey W. Rayner-Canham, 215-217. Montreal: McGill-Queen's University Press, 1997.

Hallowes, David. *How Corporations Rule: part 1: SASOL and South Africa's climate policy*. Friends of the Earth, December 2011. https://www.foei.org/wp-content/uploads/2014/01/Sasol-Report-2.pdf.

Hamilton, Lisa Anne and Steven Felt. *Plastic & Climate: The Hidden Costs of a Plastic Planet.* Center for International Environmental Law, May 2019. https://www.ciel.org/wp-content/uploads/2019/05/Plastic-and-Climate-FINAL-2019.pdf.

Hamilton, Terry. *Report: A Visual Description of the Concrete Exterior of the Cactus Crater Containment Structure*. Lawrence Livermore National Laboratory, October 2013. https://marshallislands.llnl.gov/ccc/Hamilton_LLNL-TR-648143_final.pdf.

Han, Simon. "The Impossible Task of Remembering the Nanking Massacre." *The Atlantic*, December 17, 2017. https://www.theatlantic.com/international/archive/2017/12/nanking-china-japan-iris-chang/548516/.

Hargittai, István. *Judging Edward Teller: A Closer Look at One of the Most Influential Scientists of the Twentieth Century*. Amherst, New York: Prometheus Books, 2010.

Harman, Graham. *Heidegger Explained: From Phenomenon to Thing*. Chicago, IL: Open Court Publishing, 2007.

Hartman, Saidiya. *Lose Your Mother: A Journey Along the Atlantic Slave Route*. New York: Farrar, Straus and Giroux, 2008.

Heidegger, Martin. *Bremen and Freiberg Lectures: Insight into That Which Is and Basic Principles of Thinking*. Bloomington, IN: Indiana University Press, 2012.

——. *Introduction to Metaphysics*. Translated by Gregory Fried and Richard Polt. New Haven: Yale University Press, 2000.

——. "The Question Concerning Technology." In *The Question Concerning Technology and Other Essays*. Translated by William Lovitt. New York: Harper Perennial, 1977.

——. "The Thing." In *Poetry, Language, Thought*. Translated by Albert Hofstadter. New York: Harper & Row, 1971.

Helfand, Judith, and Daniel B. Gold. *Blue Vinyl*. 2002. https://judithhelfand.com/films/blue-vinyl/.

Hochschild, Arlie Russell. *The Managed Heart: Commercialization of Human Feeling*. 3rd ed. Oakland: University of California Press, 2012.

Hochschild, Arlie Russell. *Strangers in Their Own Land: Anger and Mourning on the American Right*. New York: The New Press, 2016.

Hoganson, Kristin L. *The Heartland: An American History*. New York: Penguin Books, 2019.

Hohn, Donovan. "Moby-Duck." *Harper's Magazine*. December 31, 2006. https://harpers.org.

——. *Moby-Duck: The True Story of 28,800 Bath Toys Lost at Sea*. New York: Penguin, 2012.

Holmes, Richard. *Coleridge: Early Visions*. New York: Viking, 1989.

Homer. *The Odyssey*. Translated by Emily Wilson. New York: W.W. Norton, 2018.

Horikoshi, Jiro. *Eagles of Mitsubishi: The Story of the Zero Fighter*. Translated by Shojiro Shindo and Harold N. Wantiez. Seattle: University of Washington Press, 1992.

Howes, Ruth H., and Caroline L. Herzenberg. *Their Day in the Sun: Women of the Manhattan Project*. Philadelphia: Temple University Press, 1999.

Huntsberry, Jawanna. "Jawanna Huntsberry Interview Session I." Interviewed by Chelsea Arsenault. Mossville History Project, LSU Libraries Special Collections Williams Center for Oral History (June 3, 2016): transcript 75-76. https://lib.lsu.edu/sites/all/files/oralhistory/presentations/mossvilledirectory/transcript/Huntsberry_Jawanna_T4624.pdf.

Jackson, Roger Clay. "Roger Clay Jackson Interview Session I." Interviewed by Chelsea Arsenault. Mossville History Project, LSU Libraries Special Collections Williams Center for Oral History (October 31, 2015): transcript 21. https://lib.lsu.edu/sites/all/files/oralhistory/presentations/mossvilledirectory/transcript/Jackson_Roger_Clay_T4535.pdf.

Jambeck, Jenna R., Roland Geyer, Chris Wilcox, Theodore R. Siegler, Miriam Perryman, Anthony Andrady, Ramani Narayan, and Kara Lavender Law. "Plastic Waste Inputs from Land into the Ocean." *Science (American Association for the Advancement of Science)* 347, no. 6223 (2015): 768-771. DOI: 10.1126/science.1260352.

Jenkins, Dennis L., Loren G. Davis, Thomas W. Stafford, Paula F. Campos, Bryan Hockett, George T. Jones, Linda Scott Cummings, Chad Yost, Thomas J. Connolly, Robert M. Yohe, Summer C. Gibbons, Maanasa Raghavan, Morten Rasmussen, Johanna L. A. Paijmans, Michael Hofreiter, Brian M. Kemp, Jodi Lynn Barta, Cara Monroe, M. Thomas P. Gilbert, and Eske Willerslev. "Clovis Age Western Stemmed Projectile Points and Human Coprolites at the Paisley Caves." *Science (American Association for the Advancement of Science)* 337, no. 6091 (2012): 223-228. DOI: 10.1126/science.1218443.

Jetñil-Kijiner, Kathy. *Iep Jāltok: Poems from a Marshallese Daughter*. Tucson: University of Arizona Press, 2017.

——. "Anointed." April 16, 2018. www.kathyjetnilkijiner.com.

Jones, Brenda Cole. "Brenda Cole Jones Interview Session I." Interviewed by Stephanie Dragoon. Mossville History Project, LSU Libraries Special Collections Williams Center for Oral History (May 6, 2015): transcript 43. https://lib.lsu.edu/sites/all/files/oralhistory/presentations/mossvilledirectory/transcript/Jones_Brenda_Cole_T4622.pdf.

Kaplan, Robert. *The Nothing That Is: A Natural History of Zero*. Oxford: Oxford University Press, 1999.

——. "What Is the Origin of Zero?" *Scientific American,* February 28, 2000. www.scientificamerican.com.

Kaufmann, Carl B. "Leo H. Baekeland." In *100+ years of Plastics: Leo Baekeland and Beyond*, edited by E. Thomas Strom and Seth C. Rasmussen. Washington D.C.: American Chemical Society, 2011. https://pubs.acs.org/doi/pdf/10.1021/bk-2011-1080.ch001.

Kawano, Yukiyo. "Active Intransitive: Memories, Histories, Places, Bodies." Unpublished graduate school paper, Vermont College of Fine Arts, 2012.

Kekulé, August. "August Kekulé Speech Berlin City Hall, 1890" in "August Kekulé and the Birth of the Structural Theory of Organic Chemistry in 1858." Translated by O. Theodor Benfey. *Journal of Chemical Education* 35. no. 1 (1958): 21. DOI: 10.1021/ed035p21.

Kimmerer, Robin Wall. *Braiding Sweetgrass.* Minneapolis, MN: Milkweed, 2013.

Knott, Richard C. *Black Cat Raiders of WWII.* Annapolis: Naval Institute Press, 1981.

Larsen, Russell D. "Kekulé's Benzolfest Speech: A Fertile Resource for the Sociology of Science." In *The Kekulé Riddle: A Challenge for Chemists and Psychologists*, edited by John H Wotiz, 177-193. Clearwater, FL: Cache River Press, 1993.

Latour, Emma Jourdan, Carol Jourdan Porter, Kathy Jourdan Jefferson, and Claudia Jourdan Handy. "Jourdan Family Interview Session I." Interviewed by Chelsea Arsenault. Mossville History Project, LSU Libraries Special Collections Williams Center for Oral History (June 10, 2016): transcript 110. https://lib.lsu.edu/sites/all/files/oralhistory/presentations/mossvilledirectory/transcript/Jourdan_Family_T4625.pdf.

Lazare, Aaron. *On Apology*. Oxford: Oxford University Press, 2004.

Lee, Kuan Ken, Mark R. Miller, and Anoop S. V. Shah. "Air Pollution and Stroke." *Journal of Stroke* 20, no. 1 (2018): 2-11. DOI: 10.5853/jos.2017.02894.

Lehrer, P. "Anger, Stress, Dysregulation Produces Wear and Tear on the Lung." *Thorax* 61, no. 10 (2006): 833-834. DOI: 10.1136/thx.2006.057182.

Light, Jennifer S. "When Computers Were Women." *Technology and Culture* 40, no. 3 (1999): 455-483. DOI: 10.1353/tech.1999.0128.

Liittschwager, David, and Susan Middleton. *Archipelago: Portraits of Life in the World's Most Remote Island Sanctuary*. Washington, D.C.: National Geographic Society, 2005.

Lorde, Audre. "The Poet as Outsider." *Dream of Europe: Selected Seminars and Interviews 1984-1992*. Chicago: Kenning Editions, 2020.

Lorde, Audre. "The Transformation of Silence Into Language and Action" and "Poetry Is Not a Luxury." In *Sister Outsider: Essays and Speeches*. Berkeley, CA: Crossing Press, 2012.

Los Alamos National Laboratory. "Fogbank, Lost Knowledge Regained." *Nuclear Weapons Journal* 2, (2009): 20-21. https://www.lanl.gov/science/weapons_journal/wj_pubs/17nwj2_09.pdf.

Los Alamos Science. "Special Issue: Stanisław Ulam 1909-1984." *Los Alamos Science* 15 (1987). https://la-science.lanl.gov/lascience15.shtml.

Loudon, Irvine. *Death in Childbirth: An International Study of Maternal Care and Maternal Mortality 1800-1950.* Oxford Clarendon Press, 1992.

Mann, Susan A. "Pioneers of U.S. Ecofeminism and Environmental Justice." *Feminist Formations* Vol. 23, No. 2 (Summer 2011).

Matsumoto, Samantha, Allison Frost, and Kanani Cortez. "Pacific Islanders Disproportionately Affected by COVID-19." *Think Out Loud*, Oregon Public Broadcasting, July 1, 2020. https://www.opb.org/radio/programs/think-out-loud/article/equitable-giving-university-research-pacific-islander-community/.

Maynard, Micheline. *The End of Detroit: How the Big Three Lost Their Grip on the American Car Market.* New York: Currency, 2004.

McDowell, Robin. "Black Resistance in Louisiana's Cancer Alley." *Boston Review,* June 4, 2019. www.bostonreview.net.

McQueen, Alison. *Empress Eugénie and the Arts: Politics and Visual Culture in the Nineteenth Century*. Burlington, VT: Ashgate, 2011.

Meigs, Charles. *Females and Their Diseases*. Philadelphia: Lea and Blanchard, 1848.

——. *On the Nature, Signs and Treatment of Childbed Fever*. Philadelphia: Lea and Blanchard, 1854.

Meikle, Jeffrey L. *American Plastic: A Cultural History*. New Brunswick, NJ: Rutgers University Press, 1995.

Meitner, Lise. "Looking Back." *Bulletin of Atomic Scientists* (November 1964).

Messimer, Dwight R. *In the Hands of Fate: The Story of Patrol Wing Ten, 8 December 1941–11 May 1942*. Annapolis: Naval Institute Press, 1985.

Middleton, Susan, and Liittschwager, David. "Hawaii's Outer Kingdom." *National Geographic* 208, no. 4 (2005): 70.

Mitter, Rana. *Forgotten Ally: China's World War 1937-1945*. New York: Mariner Books, 2013.

Morland, Howard. *The Secret That Exploded*. New York: Random House, 1981.

Morris, Peter J.T., and Anthony S. Travis. "A History of the International Dyestuff Industry." *American Dyestuff Reporter* 81, no. 11 (1992). http://www.colorantshistory.org/HistoryInternationalDyeIndustry.html

Morris, Richard, Tamara Morris, and Tatiana Osipovich. "Old Believers." *Russian-Speaking Communities in Oregon*. https://sites.google.com/a/lclark.edu/rsco/immigrant-communities/old-believers.

Mycielski, Jan. "Measurable Cardinals, Ergodicity, Biomathematics." In *From Cardinals to Chaos: Reflection on the Life and Legacy of Stanislaw Ulam*, edited by N.G. Cooper, Roger Eckhardt, Nancy Shera. Cambridge: Cambridge University Press, 1989.

Napier, Susan. *Miyazakiworld: A Life in Art*. New Haven: Yale University Press, 2018.

Nead, Lynda. "Fashion and Visual Culture in the 19th Century: The Crinoline Cage." Lecture, Gresham College, London, February 4, 2014. www.gresham.ac.uk/lectures-and-events/fashion-and-visual-culture-in-the-19th-century-the-crinoline-cage.

Ng, Teddy. "Beijing and Seoul furious at Shinzo Abe's visit to Yasukuni Shrine." *South China Morning Post,* December 26, 2013. https://www.scmp.com/news/asia/article/1390047/beijing-and-seoul-furious-shinzo-abes-visit-yasukuni-shrine.

Nijboer, Donal. *Seafire vs. A6M Zero: Pacific Theatre*. Oxford: Osprey Publishing, 2009.

Nordheim, G., H. Sponer, and E. Teller. "Note on the Ultraviolet Absorption Systems of Benzene Vapor." *The Journal of Chemical Physics* 8, no. 6 (1940). DOI: 10.1063/1.1750688.

Notley, Alice. *Close to Me & Closer...(The Language of Heaven) and Désamère.* Berkley: O Books, 1995.

——. *Culture of One.* New York: Penguin Books, 2011.

——. *Mysteries of Small Houses.* New York: Penguin, 1988.

Nuland, Sherwin B. *The Doctors' Plague: Germs, Childbed Fever, and the Strange Story of Ignác Semmelweis.* New York: W.W. Norton, 2003.

Okumiya, Masatake, Jiro Horikoshi, and Martin Caidin. *Zero!* Auckland: Pickle Partners Publishing, 2014.

Oliner, Samuel P. *Altruism, Intergroup Apology, Forgiveness, and Reconciliation.* Saint Paul, MN: Paragon House, 2008.

Paradowski, Robert J. "Linus Pauling: American Scientist." *Britannica,* accessed August 15, 2020. https://www.britannica.com/biography/Linus-Pauling/Humanitarian-activities.

Parry, Richard Lloyd. "The school beneath the wave: the unimaginable tragedy of Japan's tsunami." *The Guardian,* August 24, 2017. https://www.theguardian.com/world/2017/aug/24/the-school-beneath-the-wave-the-unimaginable-tragedy-of-japans-tsunami.

Parsons, Keith M., and Robert A. Zaballa. *Bombing the Marshall Islands: A Cold War Tragedy.* Cambridge: Cambridge University Press, 2017.

Pearson, Michael. "On the Belated Discovery of Fission." *Physics Today* 68, no. 6 (2015): 40-45. DOI: 10.1063/pt.3.2817.

Penberthy, Jeff. "Plane Designer Recalls Days of Zero's Success." *Los Angeles Times,* December 14, 1972. www.latimes.com.

Perras, Michael, and Silke Nebel. "Satellite Telemetry and its Impact on the Study of Animal Migration." *Nature Education Knowledge* 3. no. 12 (2012): 4. https://www.nature.com/scitable/knowledge/library/satellite-telemetry-and-its-impact-on-the-94842487/#:~:text=To%20conclude%2C%20the%20use%20of,the%20study%20of%20migration%20physiology.

Philips, Ari "Plastics Pollution on the Rise." Environmental Integrity Project, September 5, 2019.

Podhajny, Dr. Richard M. "History, Shellfish, Royalty, and the Color Purple." *Paper, Film and Foil Converter,* July 1, 2002. http://pffc-online.com/mag/1348-paper-history-shellfish-royalty.

Ragusa, Antonio, Alessandro Svelato, Criselda Santacroce, Piera Catalano, Valentina Notarstefano, Oliana Carnevali, Fabrizio Papa, Mauro Ciro, Antonio Rongioletti, Federico Baiocco, Simonetta Draghi, Elisabetta D'Amore, Denise Rinaldo, Maria Matta, Elisabetta Giorgini. "Plasticenta: First evidence of microplastics in human placenta." *Environment International* Volume 146, January 2021, 106274. https://doi.org/10.1016/j.envint.2020.106274

Rankine, Claudia. *Don't Let Me Be Lonely: An American Lyric*. Minneapolis: Graywolf Press, 2004.

Retallack, Joan. *The Poethical Wager*. Oakland: University of California Press, 2004.

Rhodes, Richard. *Dark Sun: The Making of the Hydrogen Bomb*. New York: Simon and Schuster, 1995.

——. *The Making of the Atomic Bomb*. New York: Simon and Schuster, 1986.

Robertson, Lisa. "7.5 Minute Talk for Eva Hesse." In *Nilling*. Toronto: Book Thug, 2012.

Rocke, Alan J. *Image and Reality: Kekulé, Kopp, and the Scientific Imagination*. Chicago: University of Chicago Press, 2010.

Rogers, Heather. *Gone Tomorrow: The Hidden Life of Garbage*. New York: The New Press, 2006.

Rollerson v. Port Freeport. 3:2018-cv-00235 (txsd). Court Docket Sheet, accessed February 9, 2020. www.docketbird.com/court-cases/Rollerson-v-Port-Freeport.

Rota, Gian-Carlo. "The Lost Café." *Los Alamos Science* 15, no. 2 (1987): 23-32. https://permalink.lanl.gov/object/tr?what=info:lanl-repo/lareport/LA-UR-87-3600-03.

Royal Society of Chemistry. "Sir William Perkin and the 150th Anniversary of the Discovery of Mauveine." Copyright 2006. https://www.rsc.org/Chemsoc/Activities/Perkin/2006/index_non_flash.html

Rust, Susanne. "How the U.S. betrayed the Marshall Islands, kindling the next nuclear disaster." *Los Angeles Times,* November 10, 2019. www.latimes.com.

——. "Radiation in parts of the Marshall Islands is far higher than Chernobyl, study says." *Los Angeles Times,* July 15, 2019. www.latimes.com.

Saaler, Sven. "Nationalism and History in Contemporary Japan." *The Asia-Pacific Journal* 14, issue 20, no. 7 (2016). https://apjjf.org/2016/20/Saaler.html.

Safina, Carl. *Eye of the Albatross*. New York: Henry Holt and Company, 2002.

Safranski, Rüdiger. *Martin Heidegger: Between Good and Evil*. Translated by Ewald Osers. Cambridge, MA: Harvard University Press, 1998.

Sandberg, Gösta. *The Red Dyes: Cochineal, Madder, and Murex Purple: A World Tour of Textile Techniques*. Translated by Edith M. Matteson. Asheville, NC: Lark Books, 1997.

Santoro, Helen. "These Tiny Microbes Are Munching Away at Plastic Waste in the Ocean." *Science (American Association for the Advancement of Science)*, May 20, 2019. DOI: 10.1126/science.aay0670.

Santos, Gildo Magalhães. "A Tale of Oblivion: Ida Noddack and the 'Universal Abundance' of Matter." *Notes and Records of the Royal Society of London* 68, no. 4 (2014): 373-389. DOI: 10.1098/rsnr.2014.0009.

Segre, Emilio. *Enrico Fermi, Physicist*. Chicago: University of Chicago Press, 1970.

Seife, Charles. *Zero: The Biography of a Dangerous Idea*. New York: Penguin Books, 2000.

Serber, Robert. *The Los Alamos Primer: The First Lectures on How to Build an Atomic Bomb*. Edited by Richard Rhodes. Oakland: University of California Press, 1992.

Seward, Desmond. *Eugénie: The Empress and Her Empire*. Stroud, UK: Sutton Publishing, 2004.

Shearman, Bill. *A Forgotten Community: A History of Mossville*. Sulphur, Louisiana: Wise Publications Printing, 2017. https://imperialcalcasieu museum.org/wp-content/uploads/2015/11/A-Forgotten-Community-for-WEBSITE.pdf.

Sime, Ruth Lewin. *Lise Meitner: A Life in Physics*. Oakland: University of California Press, 1997.

Smith, Ali. "On Offer and On Reflection." In *Artful*. New York: Penguin Books, 2013.

Smith-Norris, Martha. *Domination and Resistance: The United States and the Marshall Islands During the Cold War*. Honolulu: University of Hawai'i Press, 2016.

Snodgrass, Roger. "A not-so-mysterious woman." *Los Alamos Monitor*. August 3, 2008. www.lamonitor.com.

Sobelman, 'Annah. *The Tulip Sacrament*. Hanover, NH: Wesleyan University Press, 1995.

Solnit, Rebecca. *Recollections of My Nonexistence: A Memoir.* New York: Viking, 2020.

Spector, Tami I. "Nanoaesthetics: From the Molecular to the Machine." *Representations* 117. no. 1 (2012): 1-29. DOI: 10.1525/rep.2012.117.1.1.

Swallow, J.C. "A History of Polythene." In *Polythene*, edited by A. Renfrew and Phillip Morgan. London: Iliffe; New York: Interscience Publishers, 1960.

Taylor, Dorceta E. *The Environment and the People in American Cities, 1600s-1900s: Disorder, Inequality, and Social Change*. Durham, NC: Duke University Press, 2009.

——. *Toxic Communities: Environmental Racism, Industrial Pollution, and Residential Mobility*. New York: NYU Press, 2014.

——. *The Rise of the American Conservation Movement: Power, Privilege, and Environmental Protection*. Durham, NC: Duke University Press, 2016.

Texas Department of State Health Services. "Assessment of the Occurrence of Cancer, Freeport, Texas, 2000-2015." May 2, 2018.

Tuck, Eve. "Suspending Damage, a Letter to Communities." Harvard Educational Review Vol. 79 No. 3 (2009): 409-427.

Ulam, Adam B. *Understanding the Cold War: A Historian's Personal Reflections*. New Brunswick, NJ: Transaction Books, 2002.

Ulam, Alex. "Lviv's, and a Family's, Stories in Architecture." *New York Times,* October 17, 2013. https://www.nytimes.com/2013/10/20/travel/lvivs-and-a-familys-stories-in-architecture.html.

Ulam, Françoise. "Postscript to Adventures." Epilogue to *Adventures of a Mathematician* by Stanislaw M. Ulam. Oakland: University of California Press, 1991.

Ulam, Stanislaw M. *Adventures of a Mathematician*. Oakland: University of California Press, 1991.

UN Climate Change. "COP24 Live Conversation with Kathy Jetñil-Kijiner, a Poet and Climate Activist from the Marshall Islands." Facebook, December 8, 2019. https://www.facebook.com/watch/live/?v=2368308010166372&ref=watch_permalink.

United States Department of State. "The Washington Naval Conference, 1921-1922." Office of the Historian, Foreign Service Institute. https://history.state.gov/milestones/1921-1936/naval-conference.

Valéry, Paul. "The Graveyard by the Sea," trans. C. Day Lewis. https://www.poemhunter.com/poem/the-graveyard-by-the-sea/.

Veracini, Lorenzo. *The Settler Colonial Present*. New York: Palgrave Macmillan, 2015.

Walsh, Julianne M., Hilda C. Heine, Carmen Milne Bigler, and Mark Stege. *Etto ñan Raan Kein: A Marshall Islands History*. Honolulu: Bess Press, 2012.

War Department. "You're Going to Employ Women." Washington D.C., 1943. https://massachusettsarchives.files.wordpress.com/2015/12/youre-going-to-employ-women-pamphlet.pdf.

Watkins, Sallie A. "Lise Meitner, the Foiled Nobelist." In *A Devotion to Their Science: Pioneer Women of Radioactivity*, edited by Marlene F. Rayner-Canham and Geoffrey W. Rayner-Canham, 217. Montreal: McGill-Queen's University Press, 1997.

Weiss, Kenneth R. "Altered Oceans: Part Four: Plague of Plastic Chokes the Seas." *Los Angeles Times*, August 2, 2006. https://www.latimes.com/world/la-me-ocean2aug02-story.html#:~:text=An%20estimated%201%20million%20seabirds,turtles%20suffer%20the%20same%20fate.&text=The%20amount%20of%20plastic%20in,risen%20sharply%-20since%20the%201950s.

Wertz, Dorothy, and Richard Wertz. *Lying In: A History of Childbirth in America*. New Haven, CT: Yale University Press, 1989.

Whyman, Tom. "The ghosts of our lives: From communism to dubstep, our politics and culture have been haunted by the spectres of futures that never came to pass." *New Statesman,* July 31, 2019. https://www.newstatesman.com/politics/uk/2019/07/ghosts-our-lives.

Wilkinson, Stephan. "Myth of the Zero," *Aviation History Magazine.* December 1, 2017, www.historynet.com.

Wilson, Nicholas Hoover, and Brian Jacob Lande. "Feeling Capitalism: A Conversation with Arlie Hochschild." *Journal of Consumer Culture* 5, no. 3 (2005): 275-288. DOI: 10.1177/1469540505056789.

Wright, Stephanie L., Darren Rowe, Richard C. Thompson, and Tamara S. Galloway. "Microplastic Ingestion Decreases Energy Reserves in Marine Worms." *Current Biology* 23, no. 23 (2013): R1031-R1033. DOI: 10.1016/j.cub.2013.10.068.

Yoshida, Takashi. *The Making of the "Rape of Nanking": History and Memory in Japan, China, and the United States*. Oxford: Oxford University Press, 2006.

Yoshimura, Akira. *Zero Fighter.* Translated by Retsu Kaiho and Michael Gregson. Westport, CT: Praeger Publishers, 1996.

Yusoff, Kathryn. *A Billion Black Anthropocenes or None*. Minneapolis: University of Minnesota Press, 2018.

ACKNOWLEDGMENTS

My investigations in this book are inexpert. I am a writer and poet, not a historian, sociologist, or scientist. My status as inexpert is a fact, and also a mode: The inexpert investigation involves swerves and leaps, unlikely connections, fortuitous juxtapositions. It allows, to quote Joan Retallack: "the necessarily inefficient, methodically haphazard inquiry characteristic of actually living with ideas." This mode does not suggest, however, that "anything goes," as Jules Boykoff and Kaia Sand write in their essay, "Notes on inexpert investigation." Rather, the inexpert inquiry entails response-ability—a living response and practice of accountability to the lives and histories encountered. The inexpert also invites response from others. It rejects the idea that anything is the "final word." It is a way of stepping into an endlessly entangled and ongoing conversation, conducted through archives, stories, artifacts, actions, and lives.

I'm humbled with gratitude for all who participated with me in the entangled conversations of this book. I'm grateful to my family: my father, Don Cobb, and my late mother, Connie; my partner Cobalt Coy, Aster Hartman-Coy, Jen Coleman, Paula Ward, and Sue Landers. Each gave nurture and care, and hours of their lives: talking with me, traveling with me, tolerating plastic trash, reading the manuscript, and guiding it as it developed. Sue edited a first full draft with her usual precision, intellect, and love. I could not have written this without my beloved friend Yukiyo Kawano; our collaborations and conversations make everything I do better. Dominique Browning was a hero of this book over ten years, encouraging, guiding, and challenging. Her edit of the manuscript and her input focused and sharpened the project and made it stronger. The idea for the book arose out of conversations in 2010 and 2011 with the poets and activists Alicia Cohen and Kaia Sand. CA Conrad and Brian Teare have given years of friendship, support, and encouragement.

JD Pluecker hosted me in their home, read the manuscript, and reminded me about joy and other important issues. They also connected me to others who became entangled with the book. Jules Boykoff read the manuscript and shared insights and guidance, as did Lynn Keller, Stephen Motika, Craig Santos Perez, Heather Toney, and Irene Vázquez. I'm deeply grateful to Stephen for agreeing to publish this book. Lindsey Boldt edited the manuscript for Nightboat and zeroed in on all the rough spots. Gina Kelley formatted the references and created the bibliography. Jacob Kahn copyedited the book with precision and wisdom—I'm grateful to him for preventing several errors.

Many people welcomed me into their lives. Noni and Ron Sanford took me to Kamilo Point and hosted me in their home, and Noni shared her plastic obsession with me. Mattie Larson and Bill Gilmartin of the Hawai'i Wildlife Fund also brought me to Kamilo and graced me with their knowledge and their passion for the islands. Susan Middleton hosted me at her office in San Francisco and shared stories of Shed Bird. Melanie Oldham welcomed me to Freeport, Texas, introduced me to Jessie Parker and Manning Rollerson, and generously gave time and information to me over the years. Jessie and Manning also shared their time, and Jessie became a dear friend and mentor. Errol Summerlin welcomed my family with open arms, gave us a tour of Portland, Texas, and helped me understand the threats facing his beloved community and the larger Corpus Christi region. Isabel Araiza shared information about herself and the history of Corpus Christi. Kianna Judah-Angelo invited me into her home and described her experience of coming to know her Marshallese past, and her worries about the future. Kathy Jetñil-Kijiner shared insights with me about the Marshall Islands.

Others provided timely input, support, and guidance. Andrew Hutson offered advice for tracing the plastic car part. Richard Denison gave information about the car part's plastic composition. Rebecca Altman generously shared her expertise and passion about plastic. Claire Weiner spoke to me about her father, Stan Ulam. The Naval historian Louis Dorny shared information about Elwyn Christman, and about the potential fate of the piece of plastic from Shed Bird. Philip Bassett sent his uncle Joseph F. Long's WWII journal entries. The late Raymond Hurlbut talked to me about serving with Christman. Lance Christman generously shared stories, photos, and letters about the father he never met. Curtis Ebbesmeyer introduced me to Noni Sanford and deepened my knowledge of oceanic gyres. Michelle Boyd advised me on sociological practice and oral histories. Louise Hanna and Bob Kraus hosted me twice, in Los Alamos and in Hawai'i. The attorney Amy Dinn shared her insights about Freeport's East End. Robert Hodge and Julia Barbosa Landois spoke with me about their art. Yudith Nieto told me about Mossville. The legendary Wilma Subra, who devotes her energy to helping so many, talked to me about Mossville and toxic threats facing communities throughout Louisiana. Tami Spector spoke with me about molecules and chemistry, and became a friend. Richard Rhodes provided generous details on the use of plastic in the hydrogen bomb. Susan Schultz arranged the hike to see my first albatross and tipped me off to the violent attack on the birds in Oahu. My aunt Pat Salmons shared stories about my grandfather. Grace Lewis told me about the *Plastics Pollution on the Rise* report. Arlie Hochschild connected me to Michael Tritico, who provided information and maps of the Mossville area.

Rodrigo Toscano educated me about labor movements. Sandra Phillips and Stephen Vincent drew important connections for me, as did Allyn West. Brenda Fowler and Meghan Maris were wise teachers. Meigra Simon introduced me to the photographer Jerry Takigawa. Siegfried Hecker connected me with Richard Rhodes. Karen Barad, Jeffrey Meikle, and Lauret Savoy took an interest and gave encouragement.

I learned to write, and to think, from other writers. Audre Lorde, Alice Notley, and Claudia Rankine sit at the top of my pantheon. From Richard Rhodes and Jeffrey Meikle, I learned to make compelling meaning without sacrificing historical complexity. I could not have written this book without their scholarship.

Over the years, several people taught sections of the book and invited me to speak and give workshops and readings about plastic. This helped me keep the project alive. Among them: David Abel, K. Lorraine Graham, Paolo Javier, Pattie McCarthy, Miranda Mellis, Jena Osman, Jen Scappettone, Susan Schultz, Brian Teare, and Stacey Tran.

Sections of the book appeared in *Aufgabe*, *C-L Newsletter*, *The Poetry Project Newsletter*, *Talisman,* and *Volta*. I'm grateful to the editors, and in particular to Andy Fitch, who published an excerpt of the book online as a chapbook on Essay Press, which helped give the project shape, momentum, and life at a critical time when it was flagging. Daniela Molnar and I collaborated on a walk and talk about plastic for her "Words in Place" project. I collaborated with the filmmaker Jodie Cavalier and the dancer Allie Hankins on a performance based on the book for Stacey Tran's Pure Surface series. Kaia Sand and I performed together for David Abel's "The Last Glacier" show at Passages Books. Kaia made me a skirt with the word "desire" burned into it.

The Djerassi Resident Artists Program and the Playa Art and Science Residency provided time and space to write.

So many over the years have broadened my understanding and enriched my existence; I know I have left some out of this accounting. Please forgive my faulty memory and record keeping and know that you have my gratitude for the imprint you made. The shortcomings, blindnesses, and errors in this book all come from my own limitations; I consider it the project and adventure of a lifetime to keep trying to step outside of them.

NIGHTBOAT BOOKS

Nightboat Books, a nonprofit organization, seeks to develop audiences for writers whose work resists convention and transcends boundaries. We publish books rich with poignancy, intelligence, and risk. Please visit nightboat.org to learn about our titles and how you can support our future publications.

The following individuals have supported the publication of this book. We thank them for their generosity and commitment to the mission of Nightboat Books:

Kazim Ali
Anonymous (4)
Jean C. Ballantyne
The Robert C. Brooks Revocable Trust
Amanda Greenberger
Rachel Lithgow
Anne Marie Macari
Elizabeth Madans
Elizabeth Motika
Thomas Shardlow
Benjamin Taylor
Jerrie Whitfield & Richard Motika

In addition, this book has been made possible, in part, by grants from the National Endowment for the Arts, the New York City Department of Cultural Affairs in partnership with the City Council, and the New York State Council on the Arts Literature Program.

ALLISON COBB is the author of *After We All Died*, *Born2*, and *Green-Wood*. Cobb's work has appeared in *Best American Poetry*, *Denver Quarterly*, *Colorado Review*, and many other journals. She was a finalist for the Oregon Book Award and National Poetry Series; has been a resident artist at Djerassi and Playa; and received fellowships from the Oregon Arts Commission, the Regional Arts and Culture Council, and the New York Foundation for the Arts. She lives in Portland, Oregon.